Groundwater Technology Handbook

Groundwater Technology Handbook

A field guide to extraction and usage of groundwater

A.R. MAHENDRA

Formerly of the Groundwater Exploration and
Engineering Geology Divisions,
GEOLOGICAL SURVEY OF INDIA,
Officer-in-Charge, Engineering Geology Division, SR

A comprehensive, yet concise handbook dealing with occurrence, exploration, drilling, extraction, and use of groundwater. Includes types of wells, pumps, pump selection, testing of wells, installation, piping systems, composition and purification, with figures and tables to compute yield, and flow in conduits, and troubleshooting wells.

PARTRIDGE
A Penguin Random House Company

ISBN:	Hardcover	978-1-4828-1264-0
	Softcover	978-1-4828-1265-7
	Ebook	978-1-4828-1263-3

To order additional copies of this book, contact
Partridge India
000 800 10062 62
www.partridgepublishing.com/india
orders.india@partridgepublishing.com

ACKNOWLEDGEMENTS

I am highly indebted to V.A.B.Shastri, Director, Central Ground Water Board, Government of India, for his constant guidance in scripting and reviewing the Book.

I wish to acknowledge with sincere thanks the help rendered by Praveen Kumar, Mala Raman, and Paul Danam for their unstinted support and encouragement in writing the Book, and in formatting the text and tables. My thanks are also due to T.Chenna Keshavulu for verifying the complex tables.

CONTENTS

TABLE INDEX

INDEX TO FIGURES

PREFACE

Progressive systems of exploration and extraction of groundwater have transformed desolate, inhospitable land to life sustaining tracts, providing a new meaning of life to humans and animals alike. In large parts of the Earth, groundwater is often the major source, and frequently a solitary source of water, which sustains life in isolated locales. Groundwater has been largely responsible for converting otherwise dry, harsh terrain to cultivable, hospitable land; has facilitated setting up of numerous industries in far-flung, inaccessible tracts. It has also made possible spreading the residential colonies outward and away from the City centers.

Groundwater is becoming an increasingly crucial component in estimation of the world's freshwater reserves. Relative to the land surface, groundwater has a seemingly ubiquitous presence, though it occurs at extensively unstable depths, and in indefinite quantities. Exploration and extraction of groundwater is a multi-disciplinary subject involving technology from various engineering and physical sciences. Groundwater has certain attributes, which makes it superior to surface water; significant of which are: groundwater is an economic resource; occurs in far flung areas where there are no surface water resources, is less vulnerable to pollution, is usually of high bacteriological purity, and being free of pathogenic organisms has a natural health advantage over untreated surface water.

Unlike other natural resources, groundwater is replenished by rainfall, or recharged from surface water sources. However, those who depend on groundwater pumped from a well field, either for industrial or agricultural sustenance should be responsive to certain factors that are significant to the long-term well-being of the water source. Important are:

1) Whether the wells are being over-pumped, or under-pumped;
2) The quantity of water being drawn, and its outlay;

3) Whether there is sufficient water underground, with optimum annual replenishment, to sustain the underground reservoir;
4) Whether the pumping and transmission systems installed are economical for that particular hydrological situation, and other such similar parameters, which all collate with each other.

The Book attempts, in simplistic language, to link the gap between theory and practice; between the principles of groundwater as a natural science and its application technology. Data has been compiled from literature scattered over numerous publications, and integrated with the extensive field experience of the author and his colleagues, incorporating the core material, essential for groundwater development. Emphasis is on the applied attributes of the subject.

The work is a handy reference for Hydrogeologists, Civil Engineers, Drilling Engineers, and Agriculturists who need to access various derivatives while engaged in groundwater exploration, extraction, and its use. The book attempts to elucidate the significance of procedures to coordinate the various aspects of abstracting groundwater. Each chapter is a subject by itself, on which there are numerous books.

Some of the important publications, from which the data has been compiled for the handbook are listed at the end, for further reading. The subject is so vast that it is not possible to cover all the aspects of groundwater technology; only information most frequently needed in groundwater use is included. The author and Publishers would gratefully welcome suggestions for inclusion of additional material and/or modifications in the text. The Handbook highlights and emphasizes the essential points that are fundamental to using groundwater either for domestic, industrial, or agricultural use.

armahendra@yahoo.co.in

REVIEW

This Book is the outcome of field and management experiences in groundwater exploration and application technology gathered by the Author over a period of 40 years. The author is a Geoscientist of long standing reputation, having served in the capacities of a Senior Geologist, and Head of a Division in the Geological Survey of India, and subsequently as a leading Consultant in groundwater exploration and technology. While in the groundwater division of the Geological Survey of India, he was seconded to the Technical Cooperation Mission, under the aegis of the United States Geological Survey. During the course of this assignment, he worked as a team member, in almost all the States of India, participating in a coast-to-coast groundwater exploration mission, involving advanced groundwater studies allied to deep drilling, groundwater management, and scientific usage. Later, he was Chief of a Groundwater Consultancy Firm that was awarded contracts for exploration, drilling, and development of groundwater by Irrigation and Agro Industries Development Corporations of various State governments. He was also Chief Geologist for Gulf Mining & Exploration Co., Tanzania.

The fifteen chapters in the book concisely deal with groundwater occurrence, its accretion, determination of flow parameters, flow measurements, well construction, importance of proper pump selection, and ensuring water quality for various applications. A host of easily readable tables and illustrations make the data understandable to the lay reader and the accompanying explanations meaningful to those who deal with groundwater in their day-to-day activities.

Sd/- *V.A.B. Shastri*
DIRECTOR (Retd.)
Central Groundwater Board
Government of India

INTRODUCTION

Geohydrology, as a branch of the Earth Sciences has been making rapid advances, stimulated in large measure by the increasing population, irrigation requirement to augment food production, rapid industrialization, and social survival. Groundwater is becoming an increasingly crucial component of the world's freshwater sources, that well investigated groundwater regions are becoming included as part of freshwater reserves. It is nature's paradox that the Earth's most abundant substance, water, is also the most precious; the search for which has been continuous and never ending since the dawn of civilization. It is a paradox that water occurs on Earth in abundant quantities, but the water that is available to man is an insignificant part of the total. World's seas and oceans hold about 97% of all the water on the planet. The glaciers and the polar ice caps hold about 2.25% of the remaining water in a permanently frozen, locked up state. The residual fresh water is 0.75%. Out of this fraction, only 0.11% is accessible from rivers and freshwater lakes. The remaining 0.64% forming the major portion of freshwater on earth occurs underground as groundwater.

Caspian Sea holds 75% of all the salt water in inland seas. Lakes Baikal, Tanganyika, Nyssa, Victoria and Superior hold about 60% of all the fresh water in inland lakes. The recent phenomenon of global warming, and resulting melting of the ice caps is a transient phase, which however has reduced the quantum of fresh water held by the ice shelves.

By various projections and abstractions of the measurable entities, it has been estimated that total volume of water present on the planet is 1380 million km^3, out of which only 10 million km^3 is fresh water. Against this quantum, at any point of time 1,25,000km^3 is accessible from rivers and lakes, while 8,45,000km^3 occurs under the ground as groundwater. Freshwater is replenished from annual rainfall which averages 1,10,000 km^3 over the continents. However, distribution of rainfall varies widely, causing life-threatening floods at some places, while causing critical drought and famine situations at other places.

The UN Agencies have been warning that by 2025, at least 1800 million people would be living in regions, including India, facing extreme water scarcity, and two-thirds of the world's population would be living under severe water stress. Currently, one in every six people worldwide does not have access to safe water. Surveys have shown that about 1.5 billion people live in regions where water is being over drawn depleting reservoirs and groundwater. The current hydrological cycle has to sustain an additional 2.7 billion population by 2050.

Groundwater has been accumulating over several million years, with rainfall each year adding only miniscule quantities to its static reserves. Lack of rainfall for a year or two does not intrinsically alter the total volume of groundwater in storage, though it will affect the overlying dynamic reserves. Since groundwater occurs, hidden from view, its reliability as a dependable source is often viewed with skepticism. The continuously evolving technology in groundwater abstraction is progressively demonstrating this concept as a myth.

Locating a source point, and drilling a well, is an initial and preliminary part of developing a groundwater resource. More important is the maintenance of the water source point, testing it for its potential and installation of an appropriate pumping system, as well as conveying the water economically, with minimum loss of water and power to the point of usage.

CONVERSION FACTORS:

Length:	
1 kilometer (km)	= 1000 m = 0.62 miles
1 meter (m)	= 3.28 feet (ft) = 39.37 inches
1 centimeter (cm)	= 0.01 m = 0.3937 inches
Area:	
1 sq km (km^2)	= 0.386 sq miles = 247 acres = 100 ha
1 sq meter (m^2)	= 1.196 sq yards (yds^2) = 10.744 sq feet (ft^2)
1 hectare (ha)	= 10000 sq meters = 2.471 acres
Volume:	
1 cu meter (m^3)	= 1000 liters = 220 imp gallons
	= 35.314 cu ft (ft^3)
1 cu cm (cm^3)	= 0.061 cu inches ($inches^3$)
1 liter (l)	= 0.264 US gallons = 1000 cu cms
1 cu ft(ft^3)	= 6.2 imp gallons = 28.3 liters
1 imp gal	= 4.55 liters = 1.2 US gallons
1 US gal	= 3.8 liters = 0.83 imp gallons
1 liter	= 0.22 imp gallon
Weight:	
1 kilogram (kg)	= 1000 gms = 2.205 pounds
1 gram (g)	= 0.035 ounces
Pressure:	
1 kg (force) / cm^2	= 14.223 pounds / $inch^2$
1 newton/sq.m(N/m^2)	= 0.000145 pounds / $inch^2$
1 atmosphere (atm)	= 1.033 kgf /cm^2 = 14.70 pounds / $inch^2$
Specific capacity:	= 1 imp gallon /min / ft drawdown
	= 15 liters /min / 1 m drawdown
Transmissivity:	= 1 m^2 / day = 80.5 US gallons / day/ foot = 10.74 ft^2/ day.
	= 1 US gallon/day / ft = 0.0148 m^2 / day
Permeability:	= 1 m^2/day = 20.44 imp gallons/day/ft^2
	= 24.54 US gals/day/ft^2

To obtain:

Circumference of a circle	= Diameter × π (3.1416)
Diameter of a circle	= Circumference × 0.31831
Area of a circle	= Diameter^2 × 0.7854
OR	= Diameter / 2 × Radius / 2
Surface of a sphere	= Diameter^2 × π

CONVERSION TABLES

Table 1: Volume

Units	Equivalent						
	Cu. cms	**Cu. inches**	**Litres**	**US gallons**	**Imp gal**	**Cu. feet**	**Cu. meters**
1 Cu. cm	1	0.06102	.001	0.0002642	0.0002201	0.0000353	0.000001
1 Cu. inch	16.39	1	0.016387	0.004329	0.003607	0.0005787	0.000016
1 Liter	1000	61.0234	1	0.26417	0.22008	0.03531	0.001
1 US gallon	3785.4	231	3.7854	1	0.83311	0.13368	0.003785
1 Imp gal	4542.5	277.274	4.5425	1.200	1	0.16054	0.00454
1 Cu. foot	28317	1728	28.317	7.4805	6.2321	1	0.02832
1 Cu. meter	1000000	61.023	1000	264.17	220.083	35.3145	1

Table 2: Flow

Units	Equivalent							
	Litres			Imperial gallons		US gallons		Cu. sec
	Per second	Per min	Per hour	Per min	Per hour	Per min	Per hour	
1 Litre/sec	1		3600	13.20	792	15.85	951	0.03531
1 Litre/min	0.0166	1	60	0.22	13.20	0.26	15.85	0.0005885
1 Imp gal/min	0.0757	4.54	272.76	1	60	1.20	72	0.0026757
1 Imp gal/hr	0.0012	0.075	4.54	0.016	1	0.02	1.20	0.0000446
1 US gal/min	0.063	3.785	227.0	0.833	49.96	1	60	0.00222
1 US gal/hr	0.0010	0.063	3.7854	0.0138	0.8331	0.016	1	0.000037
1 Cu. sec	28.317	1699	101941	373.93	22423.8	448.8	26928	1
1 Cum.sec	1000	60000	3600000	13198	791889	15850	951019	35.3145

Table 3: Length

Unit	Equivalent				
	centimeters	**inches**	**feet**	**yards**	**meters**
1 cm	1	0.3937	0.0328	0.01093	0.01
1 inch	2.54	1. 00	0.08333	0.0278	0.02540
1 foot	30.48	12.00	1.00	0.3333	0.3048
1 yard	91.44	36.00	3.00	1.00	0.91440
1 meter	100	39.37	3.2808	1.0936	1.00
1 km	100000	39370	3208.8	1093.6	1000
1 mile	160935	63360	5280	1760	1609.3

Table 4: Area

Unit	Equivalent				
	cm^2	**$inches^2$**	**$feet^2$**	**$yards^2$**	**$meters^2$**
1 cm^2	1.00	0.155	0.001076	0.0001196	0.0001
1 $inch^2$	6.452	1.00	0.00694	0.0007716	0.0006452
1 $foot^2$	929	144	1.00	0.1111	0.0929
1 $yard^2$	8361	1296	9.00	1.00	0.8361
1 $meter^2$	10000	1550	10.76	1.196	1
1 $mile^2$			27878400	3097600	2589998
1 acre	40465284	6272640	43560	4840	4047

ABBREVIATIONS

APT	=	Aquifer performance test
BHP	=	Brake Horse Power
cms	=	Centimeters
cu sec	=	Cubic feet per second
cum.sec	=	Cubic meters per second
GPD	=	Imperial gallons per day
GPH	=	Imperial gallons per hour
HP	=	Horse Power
Kg/cm^2	=	Kilograms per square centimeter
$Lbs/inch^2$	=	Pounds per sq inch
LPD	=	Liters per day
LPH	=	Liters per hour
LPM/ lpm	=	Liters per minute
LPS	=	Liters per second
m/mts	=	Meters
ng	=	nanogram=One billionth (10^{-9}) of a gram.
ppm	=	Parts per million
RPM	=	Revolutions per minute
Stg	=	Number of stages in the pump
TDS	=	Total dissolved solids
USGPH	=	United States gallons per hour

CHAPTER I

OCCURRENCE OF GROUNDWATER

Groundwater has a vast distribution all over the globe, but very few underground reservoirs have been well investigated, and data is scanty even for those reservoirs that have been studied.

Hydrologic cycle:

Hydrology is the study of the occurrence, movement, distribution, and circulation of water through the hydrologic cycle. Hydrologic cycle is unending; it has no beginning, and no end. Water from the oceans evaporates into the atmosphere, precipitates into atmospheric moisture, and condenses into clouds, which under certain temperature and pressure conditions result in rainfall (Fig 1). A part of this rainfall runs off the ground resulting in streams and rivers, while a part infiltrates into the ground, and drains to large underground reservoirs. The elements that coalesced, and condensed into water during the formation of the primordial earth, 4500 million years ago, is the same water that is present on earth today, and will be the same that would be present a billion years from now. No original water is being either created nor is any that already exists being destroyed. Same water continues to circulate in one form or the other.

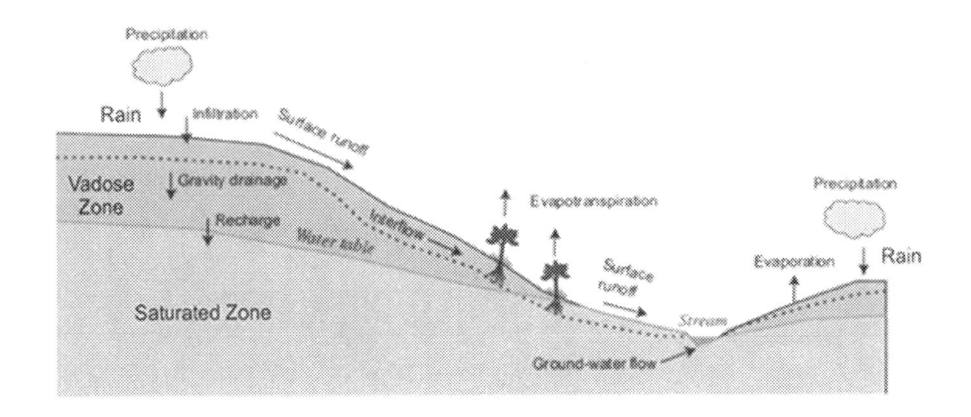

Fig 1: Graphical representation of hydrologic cycle

Groundwater occurrence, though seemingly pervasive, varies widely with depth and extent. Depth to groundwater incidence may range from 1 meter to as much as 1000 meters, dependent on the nature of the geological formations holding the water and proximity to recharge areas. Water under the ground does not occur either as streams or as underground channels.

Sub-surface water or water underground has a universal ubiquitous presence, but not all water under the ground is groundwater. Air inside any excavated area, or a drill hole is at ambient atmospheric pressure (1 kg /cm^2) and any water occurring in the interstices surrounding the well, or the drill hole should be under pressures exceeding atmospheric pressure to be able to flow into the open area. Hence, what differentiates groundwater from the sub-soil water is, groundwater occurs under pressures greater than that of the atmosphere, and that it flows freely into the excavated area. The subsurface zone of soil or rock that hold water under pressures less than the atmospheric pressure is the vadose zone, which extend below the land surface down to the water table below. Pore spaces in the vadose zone are partly filled with water and partly with air. Water in the vadose zone is held under osmotic pressure, and osmotic processes and capillary action control its movement. Vadose zones may be of various thicknesses, but in hard rock areas generally extend, on an average down to a depth of about 20 meters. Water in these zones will not sustain long term pumping. Dug wells generally tap into the vadose zones. Boreholes drilled from the floor of the dug wells interconnect the vadose and the underlying phreatic zones yielding limited quantities of water. Water levels in the vadose zone fluctuate with rainfall, and often hold perched water tables.

Underlying the vadose zone is the phreatic zone or the saturated zone in which all interconnected openings within the geologic medium are completely filled with water. Many hydro geologists separate this zone into two subzones: the phreatic zone and the capillary fringe.

In the phreatic zone interstitial water flows freely through the pores in the geologic formation, under the influence of gravity and the superincumbent pressure exerted by overlying formations. Water in the pores of the phreatic zone occurs at pressures greater than the atmospheric pressure.Lying above, and separated from the phreatic zone by the water table, is the capillary fringe. Capillary action within the voids of the geologic medium causes water to be drawn upward from the top of the phreatic zone or captured as it percolates downward from the overlying unsaturated zone. Unlike the phreatic zone, however, the capillary action causes the water in the pores to have a pressure that is lower than the atmospheric pressure. While the pores of both subzones are saturated, the different pressures in each cause the water to behave differently. Water within the phreatic zone will readily flow out of the pores while the negative pressures within the capillary fringe tightly hold the water in place. It is water from the phreatic zone that is collected and pumped from wells, and that flow into streams and springs. Water under the ground is never stationary for long at one situation, but keeps moving constantly. Gravity and the frictional forces that come into play during its movement control its passage. Rates of groundwater movement within the saturated zone may range from a few meters per year to several kilometers per day depending upon the slope of the land, hydraulic gradient and permeable nature of the formations. In certain cavernous areas, karst systems and highly faulted and fractured zones, it approaches velocities in surface flows.

The saturated zone extends downward from the capillary fringe to the depth where rock densities increase to the point when migration of fluids is no more possible. In hard rock areas, this occlusion zone may occur between 250 and 300 m, basaltic regions being an exception where water may occur even at great depths in the inter-trappean sediments. In deep sedimentary basins, this may occur at depths of approximately 5000 meters. At these extreme depths, the voids are no longer inter connected.

In soft rocks and in highly weathered and peneplained regions it generally occurs in stratum of various thicknesses, spread over a vast area and extending to indefinite depths. Sometimes they constitute huge "underground reservoirs" with an infinite supply of water. An example is the Miocene sandstones holding indeterminate quantities of fresh water under pressure, extending from Cuddalore in Tamil Nadu, via Neyveli to Rajahmundry in Andhra Pradesh. However, continuous pumping from the basin has resulted in reducing pressures, indicating approach of over-draft situation. Other similar examples in India are the Indo-gangetic-brahmaputra alluvium which hold millions of liters in underground reservoirs, and yield copious quantities of water.

Amongst the known underground reservoirs is the Nubian Aquifer, which feeds the fabled oases of Egypt and Libya. (Felicity Barringer, New York Times, 21 Nov 2011). The Nubian aquifer stretches languidly below a vast desert in northern Africa, manifested by oases and an undefined collection of water pools, migrating ever so slowly, through rock and sand to the Mediterranean sea. Carbon 14 dating has taken the researchers 50,000 years back, but isotope studies have indicated that the aquifer is probably more than a million years old. It stretches over an area of 770,000 square miles (1 square mile = 2.590 square kilometers) under the Sub-Saharan Africa. Parts of the aquifer lie 3000 meters below the Egyptian oases. The groundwater in this aquifer, below the Egyptian region alone is estimated to be 40,000 cubic kilometers in volume. How this vast body of water flows without a seemingly set course and how it is replenished has been a subject of intensive study by the Department of Energy, Argonne National Laboratory in Illinois. Four countries, Egypt, Libya, Chad and Sudan share these waters under an international agreement. These countries are resorting to heavy pumping in their own oases situated over the aquifer, but nothing is known about what is happening to the rest of the aquifer. Heavy pumping, without proper planning has resulted in some of the oases, including Kufra lake drying up.

Occurrence in geological formations:

Physiographic highs and lows in the form of broad shallow valleys, regional slopes, and their culminations are conducive for accumulation of groundwater. Water tends to drain towards physiographic lows, wherein it accumulates and tends to build positive high pressure. Geomorphic lows generally reflect and replicate an underground, hidden geological feature, such as a deeply weathered region that actually holds the draining water. These areas, recognizable by experienced Hydro geologists mark groundwater discharge boundaries, a fact frequently manifested by high yields from wells drilled in such locations.

Groundwater occurs in a wide range of geological formations, ranging from loose unconsolidated material like pebbles, sands and silts; semi-consolidated sedimentary rocks such as sandstones to hard rocks such as granites, basalts, massive limestones, and quartzites. However, the mobility of water and pressure within each of these formations varies widely; and the yield from boreholes drilled over these rocks is proportionate to the transmissivity characteristic of the formation in which the well is drilled. Output from wells depend also to a very great extent, on the method of drilling adopted, construction of the well, and most importantly the selection and installation of the appropriate pump for each individual well.

Occurrence in soft rocks:

Alluvial fills, and Aeolian deposits consisting of highly porous, and permeable material hold and transmit large quantities of ground water. Yield from these formations often exceed 2200 liters of water per minute. Some of the karstified limestones, though strictly cannot be classified as soft are however vastly permeable, and yield ample quantities of water.

In geohydrological terminology, the significance of soft rocks is synonymous with sedimentary formations, and is generally dealt with as rocks aggregating in an aqueous environment. Sedimentary formations consist essentially of sandstones, shale, claystones, siltstones, and related detrital deposits. These formations constitute 5 percent of the earth's crust, but hold about 95 percent of groundwater, the remaining 5 percent occurring in hard rocks. Amongst the sedimentary rocks, sandstones and allied porous formations constitute almost 50 percent of the earth's soft rock cover.

Sedimentary rocks are widely distributed, and possess excellent water yielding properties, with their porosities ranging from 5 to 30 percent, and accompanying synchronous permeability. Wells drilled in these formations sustain large yields over long periods. They store and transmit such large quantities of groundwater that the water levels in wells drilled in sandstones and allied rocks react little to lack of, or replenishment by annual rainfall.

Alternating beds of sandstones, shale, and limestones are characteristic of most sedimentary sequences, but individual beds may be so thick that deep wells drilled in certain sedimentary regions penetrate only one rock type. The other sedimentary formations, the shale, and clay stone possess relatively high porosities, but very low permeability. Their water yielding characteristics are relatively more analogous to hard rocks. Some representative characteristics and yield values are given in Table5. Although the yield values frequently reflect the water bearing properties of the penetrated aquifers, the drilling methodology adopted, and testing procedures affect the yield so much that the yield differences cannot be directly related to the aquifer characteristics.

Occurrence in hard rocks:

Hard rocks consist essentially of igneous and metamorphic rocks, which occupy nearly 20 percent of the earth's surface. At these places they either are exposed, or lie close to the surface. Major igneous rocks are the granites, and allied formations, which at many regions represent the original primordial crust of the cooling planet. They also occur as later intrusives, intruding into the older sedimentary formations, and into the earlier granitic crust. Metamorphic rocks represent the transformed equivalents of both igneous and sedimentary rocks, which have been subjected to intense pressure and temperature, resulting from tectonic movements. Next in importance are the volcanic effusives represented by basalts, and their variants. Present day manifestation is the lava flows ejected by active volcanoes.

Groundwater occurs in all these varieties of hard rocks to a limited extent. Wells drilled in these rocks are at the most unreliable yielders, and seldom sustain large yields. Water levels and the output react rapidly in response to intensity and duration of rainfall. Still large tracts in India, Australia, South Africa, and parts of America depend entirely on water occurring in hard rocks for their irrigation, and industrial sustenance.

In hard rocks groundwater occurs in an abstract, dissipated state held in the fractures and granular interstices; its movement is controlled by this system, and the weathered mantle overlying fresh rock. For groundwater to accrue in hard rocks, certain physical and geological situations need to be predisposed at the site to facilitate surface water infiltration and accumulation. Rainwater will not infiltrate into the ground unless the rocks are weathered, differentially weathered, or structurally disturbed, and the region is in the process of undergoing peneplanation. Normally, intensity of weathering gradually diminishes with depth; the effects of weathering generally ranging between 6 and 30 meters below ground level. In structurally deformed regions, fractures and joints with accompanying weathering and disintegration may extend down to 100 meters or more. In such tectonically disturbed regions groundwater occurs in considerable quantities to indefinite depths. In geologically disturbed areas, and in areas of deep weathering, or peneplanation, yield of groundwater has tended to become more intense with depth, rapidly increasing to even depths of 200m or more.

Fractures and joints constitute the fracture porosity, which governs the permeability of any particular hard formation. Fracture systems are limited in extent, but some of the fracture systems developed during the tectonic activity of the crust extends over long distances and to significant depths. Particularly the tensile open joints and shear fractures form major conduits for transmitting groundwater over long distances. Drill holes are located to intercept such horizons.

Following geological and geomorphic features are indicators of incidence of groundwater in hard rock terrain, and should be intensely investigated during any exploration programme:

1. Major and minor fracture zones, lineaments and zones of brecciation; It has been found that fractures and openings trending in different directions vary in their water yielding characteristics. Hence, while reconnoitering for occurrence of groundwater in hard rock areas, it is important to investigate the fracture systems.

2. Persistent topographic troughs and ridges that do not conform to regional trends of the geological and physiographic trends of the region;
3. Contact and unconformable proximity of two or more rock types;
4. Presence of intrusives, such as pegmatites and dykes, which expose kaolanised zones of low permeability, or intensely fractured zones of high permeability or loosely cemented granular derivatives;
5. Float or detrital material such as feldspar, graphic granites, epidote, olivine and minerals which suggest deep seated faulting;
6. Insitu outcrops of fresh rock interspersed with overburden, soil cover or murrum and weathered rock.

There is an indistinct, but often significant, grading interface between the weathered and fresh rock zones, which yields profuse quantities of water. Yield data from a large number of bore wells drilled in homogenous hard rocks, to depths ranging down to 300 m has shown that, maximum inflow into the well is obtained from within 100 m depth, and that yield decreases below this depth. Exceptions are where boreholes drill through fracture or fault zones, or formation contact interfaces. Frequently, wells which have apparently yielded water at deeper depths is often more an account of reducing air pressure impacting on the well walls, at depth, which permits groundwater to flow into the well, and not because groundwater is confined to lower levels. A geohydrological survey and the Drillers log can evaluate this. Well-defined pegmatite veins, fractured quartz veins, and allied intrusives that traverse granitic formations channel groundwater in moderate quantities. Wells drilled directly on such veins, or a geologically dissimilar intrusive, or wells drilled to intercept such veins wherever they occur in an inclined attitude have often sustained moderate yields in an otherwise dry terrain. Basic dykes and sills often intrude granites and related formations, which effectively act as subsurface barriers impounding groundwater towards upward slopes.

In basic basaltic formations, groundwater occurs in moderate quantities in amygdaloidal, and vesicular flows, intertrappean, and intratrappean sediments, and in minor quantities along joints, fractures, and flow contacts. Massive flows are devoid of any porosity. However, individual flows seldom exceed 10m thickness. Hence, to obtain significant quantities of groundwater in basaltic terrain, boreholes need to be drilled as deep as possible, averaging 200 meters depth, in an attempt to intercept as many flow horizons as possible.

Quantifying flow conditions:

Flow equations applicable to homogeneous media to evaluate yield potential have successfully been applied to hard rock aquifers. Extensive data on hydraulic conductivity and yield potential of a multiplicity of geological formations is available in a wide range of relevant publications. The data indicates a wide array. Table No.5 summarizes this data, consolidated from numerous bore wells drilled in hard rocks in various parts of India (Karanth 1987). The Table does not show any relationship between the hydraulic conductivity and yield. Hydraulic conductivity data has been computed from the technical reports of the Central Ground Water Board on certain restricted basins, while the yield data is the record of the actual discharge measurements made during aquifer performance tests conducted over wells drilled over a much larger area. However, the data is broadly consistent of the yield potential of the commonly occurring hard rocks, and should be viewed as an approximate lead to estimating probable discharge, before undertaking exploitation.

Based on hydro-geological factors, and consolidated recharge and output data from various geological formations, hydro-geological maps have been compiled which present an expansive picture of areas with similar conditions indicative of equivalent output potential.

Table 5: Hydraulic conductivity and yield potential of major formations. (after Karanth K.R.)

Formation		Hydraulic Conductivity (meters per day)		Recorded Yield (for average 6 meters drawdown)	
				Range	
		Range	Average	GPH	LPM
Granites	Weathered	0.0003-4.84	0.5	500-1000	38-76
	Semi-weathered	0.00009-56	2.45	200-1500	15-114
	Fractured	0.0001-2.46	0.33	400-800	30-60
Basalts	Weathered	0.002-59.7	9.10	800-2000	60-152
	Fractured	0.001-6.7	0.58	600-1200	45-90
	Vesicular	0.002-0.22	0.04	800-1500	60-114
Limestone Karstified		0.1-83	6.3	1500-7500	114-560
Shales (Cuddapah)		0.05-3.5	0.9	1200-2000	90-152
Sandstones(Bhandar)		0.13-1.56	0.66	1500-3000	114-228
Sandstones (Miocene)		2.6-351	89.0	10000-30000	760-2280
Piedmont (Bhabar)		117-927	350	30000-40000	2280-3000
Valley fill (Banas)		113-993	419	30000-45000	2280-3375
Valley terrace		5.2-269	68	15000-25000	1140-1875

CHAPTER II

EXPLORATION FOR GROUNDWATER

There are diverse methods of exploring for incidence of groundwater. The course and methodology adopted for a particular groundwater exploration project is largely dependent on the purpose, for which it is undertaken, and the volume requirement of water. For large-scale consumption, such as for industrial, irrigation, or town water supply scheme a detailed, appropriately sequenced investigation is crucial. For small quantity demands, such as for domestic consumption, a simple search is often sufficient. Single well points to meet home needs can be located, without elaborate instrumentation, if wells already exist in the area. A statistical survey of the tube wells in the area, their location, depth drilled, water yield from each of these wells is often sufficient to establish a feasible drilling point. In such cases, only essential caution should be to locate the new point as far away as possible from an existing well.

Water wells planned to supply water to large demand applications need a detailed investigation by an experienced Hydro geologist. Since groundwater lies hidden from view, exploration for its location is in the beginning an abstract exercise, subject to logical, objective interpretation. The investigation consists of a sequence of linked scientific procedures, both instrumented and experienced observation, each collateral to the other. Each of these steps narrows down the search to restricted potential ground from which final points for test drilling are selected.

Hydrogeological Investigations:

The principles of groundwater occurrence and accumulation, explained in Chapter I govern the broad approach to groundwater exploration.

Major tools would be a geological map, or a topographical map of suitable scale (1:25000 or 1:50000), a geocoded satellite imagery, a GPS location finder, Brunton compass, and a geological hammer. If a geological map and the satellite imagery are not readily available, then the alternative option is to organize a sketch map of the area, which need not be to scale. Regional investigations are initiated with a rapid geological reconnaissance of the area, and its environs, and the following features are marked on the sketch map: i) Exposed geological features; ii) major physiographic and structural features, highlighting the potential of these features to facilitate rain water infiltration; iii) probable directions of groundwater migration; iv) ground slope angles; v)areas of prospective water accumulation such as slope culminations; vi) physiographic recharge locations such as valleys and flat ground, vii) all existing wells in the area and the vicinity; viii) along with available data on these wells such as date drilled or excavated, depth, water yield, depth of striking water, operating pump, its capacity and all such data relevant to incidence of groundwater.

From this map and data, selected areas are short-listed for more detailed geological studies. During the advanced studies, geological features facilitating water infiltration, its transmigration and probable accumulation points are carefully noted. Emphasis is on grain size, and structural characteristics of the geological formations such as joints, fractures, and intrusives that contribute to porosity. General strike and inclination trend of all these features are carefully noted. The third step is to further shortlist the prospective areas in an effort to draw a correlation between the existing wells, if any, and the locations with apparent infiltration potential. Succeeding the geological survey, chosen locations are marked both on the map, and on the ground for detailed Geophysical surveys which are decisive for locating exploratory drilling points.

Geophysical Surveys:

Well-instrumented Geophysical surveys provide indirect evidence of groundwater occurrence at depth. The physical, chemical, and magnetic properties of soil and rock formations continuously vary from place to place, and the differences are further enhanced by the existence of sub-surface in-homogeneities. The electrical and magnetic characteristics are significantly influenced by presence or absence of saturation in the formations. The electrical conductance of a geological formation also varies in relation to its porosity, saturation, and salinity of the water it is holding or transmitting. Geophysical studies, which essentially probe these magnetic and electrical distinctiveness of these formations lead to interpretation of the incidence of groundwater. Amongst the various techniques employed, most rapid and versatile is the electrical resistivity method, which is based on the characteristic that saturated formations (water bearing) are more conductive, and that dry zones are resistive to the passage of an electrical current. Geophysical surveys to locate presence of groundwater underground are feasible only when large open and plain areas are available. Geophysical surveys are not applicable to domestic well point search, unless a minimum horizontal linear distance of 10 m is accessible to stretch the cables, which is seldom available in populated areas. Presence of underground cables, pipe lines and other metallic bodies will interfere with the transmission of the current, and will often provide misleading data.

The specific resistance of the rock or the soil is the resistance offered by a unit cube of the medium when a measured current transmits normal to the cross-sectional area. The resistivity+ of natural materials to the flow of an electrical current varies through a wide range. Massive granite has a resistivity of 106 ohm-meters, while clays saturated with salt water is highly conductive with a resistance of 1.0 ohm-meters. Fresh water has a resistivity of 50 ohm-meters, while aquifers carrying fresh water offer a resistance of 102 ohm-meters. Thus, passing an electric current of a known quantity through the medium, and measuring the potential difference between the two points would determine the apparent resistance of sub-surface formations. By this procedure, it is possible to approximate the saturation or the dryness of the medium through which the current is transmitting. A typical set-up for resistivity surveys is schematically shown in figure 2.

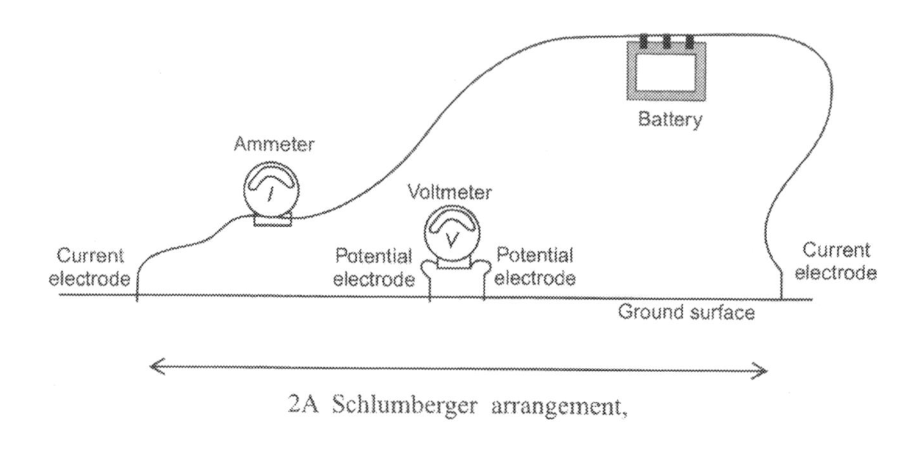

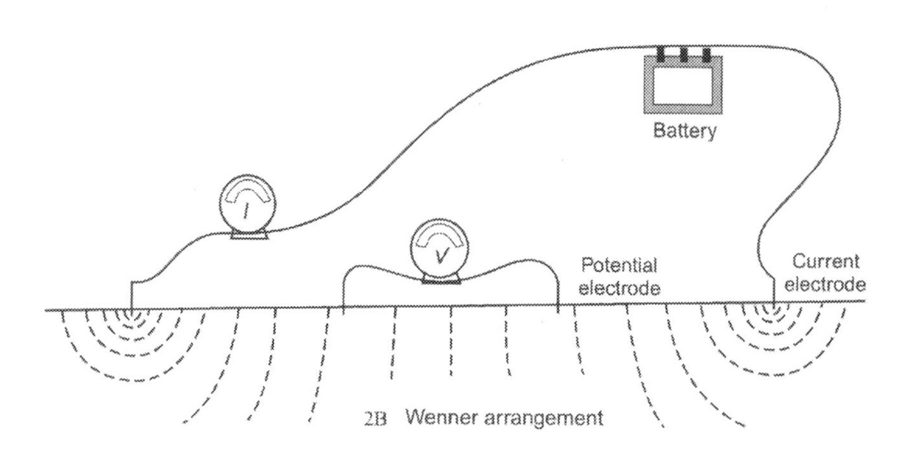

Fig 2: Diagrammatic illustration of resistivity survey for groundwater exploration

The current is applied with metal rods driven into the ground, which act as current electrodes. Distances between the electrodes vary from 10 meters to several 100 meters, depending on the depth to be probed. To preclude polarization, direct current ranging from 9 to 180 volts, (based on the depth to be explored) is applied to the current electrodes. Dry soils around these electrodes are dampened with water to ensure optimum electrical conductivity. The resulting voltage difference or the potential difference is measured with two more electrodes, called potential electrodes located symmetrically in a straight line between the two current electrodes. The potential electrodes are placed in porous ceramic pots, filled with copper sulphate solution, and are placed some distance away from the current electrodes to avoid rapid voltage fluctuation in the vicinity. Various resistivity survey procedures adopt diverse electrode configurations by shifting relative locations of current and potential electrodes. Most frequently deployed are the Schlumberger and Wenner configurations (figures 2A and 2B respectively). Power from an external battery is applied to the current electrodes via an ammeter (I), which measures the current input. The potential difference spiraling between the two inner electrodes is measures by the voltmeter (V). The ammeter and the voltmeter, with controlling toggle switches are conveniently placed in a DC Resistivity meter casing.

In the Schlumberger procedure, the potential electrodes are retained at a constant location, while the current electrodes are moved successively farther apart on either side, progressively increasing the interval "a", i.e., AB/2 (figure 2B), and measuring the resistivity values for a series of simultaneous movement of the outer electrodes. This set-up discloses a clearer interpretation of the sub-surface situation for specific outer electrode spacing, and requires less man power as the central electrodes remain stationery, while only the outer current electrodes are moved after every reading. Resistivity is computed from the formula:

$$pa = \pi \frac{(L/2)^2 - (a/2)^2}{a} \frac{V}{I}$$

Where,

pa = Apparent resistivity
V = Potential in volts measured between the two potential electrodes
I = Current applied to the ground in amps
a = Spacing between potential (inner) electrodes in meters
L = Spacing between current (outer) electrodes in meter

In the Wenner set-up the two potential electrodes are placed in-line with the current electrodes, all four being equidistant to each other, and disposed symmetrically on either side of the probe point known as the VES (Vertical Electrical Sounding) station. All the four electrodes are moved successively farther apart, away from the VES and a series of resistivity readings taken in each of the positions. The resistance varies directly as the length of the medium, and inversely as its cross-section, and is expressed by the equation:

$$pa = 2\pi a \frac{V}{I}$$

Symbols represent factors given against the preceding formula. As the electrode spacing is increased, the path of current travel beneath the surface takes a longer parabolic path, and in the process achieves a deeper penetration. It is generally assumed that the depth probed approximates the spacing of the two electrodes in case of a Wenner configuration; and is approximately equal to half the distance between the current electrodes in case of a Schlumberger procedure. However, this has been found to be not true, as the ground beneath the surface is rarely, if at all, isotropic and homogeneous. The data interpretation of depths to hidden horizontal grading boundaries, or ambiguous interfaces is made by matching the field curves with type curves that are available for hypothetical layers of specific contrasts in resistivity values. "Lows" of the field resistivity curve reflect not only saturated ground, but may also show up, consequent to presence of other conductive objects such as pockets of decomposed rock, fault zones, and even ore bearing horizons. Interpretation of resistivity curves, while searching for groundwater, become meaningful only when interpreted in close concurrence with the geology and structural conditioning of the area. It should be constantly borne in mind that geophysical surveys is a tool in the hands of the geologist conducting the exploration; and that geophysical probing is means to an end, and not an end by itself.

Remote sensing:

A versatile tool available today to the Exploration Geologist is the Remote sensing applications, which has been making rapid strides since past several years, in broad basing natural resource locations. By this technique, information on an object on the earth's surface is acquired by remote sensing satellites by scanning the electromagnetic energy emitted by the object. Variations in the reflectance properties of the objects on the earth's surface result in differing tonal expressions, called signatures, on the images recorded by on-board scanners. These satellite recorded images are interpreted by computers in the ground based receiving stations to generate credible mosaics, called imageries.

The exploration geologist, depending on the extent of the search area, has the option of selecting imagery with a range of ground resolutions and definitions. Higher the resolution more expensive is the imagery. The Indian Remote sensing satellites IRS 1 C and IRS 1 D launched in 1996 and 1997 are three axes polar orbiting sun-synchronous satellites circling the earth at altitudes of 817 kilometers. The high-resolution panchromatic sensor on board IRS 1 C has a resolution of 6 meters. Sweeping a 70 kilometer swath, the pan sensor is steerable by ground control through 52 degrees across its track enabling stereoscopic imagery, with revisit capability. The satellite also carries LISS 3 (Linear imaging self scanning sensors) which sweeps a swath path of 142 kilometers providing a ground resolution of 23.6 meters in the visible / near infra red (VNIR) range. Additionally, a short wave infrared (SWIR) range on the sensor provides a ground resolution of 70.8 meters in a 148 kilometers wide swath. A third sensor called wide field sensor sweeps a broader swath of 810 kilometers, with a spatial resolution of 88 meters, suitable for picking up large geological and geographical features. The sensors transmit synoptic data coverage over large areas in four discreet spectral bands in the visible and infrared wavelengths of 0.45 to 0.86 microns. In combination, these sensors provide enhanced spatial resolution, additional spectral bands, stereoscopic imaging, and wide field coverage. Cartostat I, also known as IRS P5 has two panchromatic cameras with spatial resolution of 2.5 meters, and a swath width of 30 kilometers each. The high resolution of the data from these cameras, with accurate imagery is being traded at the cost of multi-spectral capability, and smaller area coverage. Computer enhanced stereoscopic data yield contour maps to generate near ground like scenery of great geological value. For most regional and preliminary groundwater exploration programmes, LISS 3 and PAN merged false colored, computer enhanced imagery on 1: 50000 scale provides cost-effective and adequate base data to mark significant physiographic, geological, and structural features which merit attention in the field. The imagery when overlaid with contours from related topographical maps; and geology from the area geological maps facilitate interpretation of features conducive for groundwater accretion, thus narrowing down areas for detailed hydrogeological and geophysical investigations. Amongst the geological features, the easily recognizable are the valley fills, abandoned river channels, residual and fossil terraces, rock out-crops, and structural lineaments which control and channel groundwater flow. While using satellite imagery for groundwater exploration, it is fundamental to take into account that the imagery provide no more than a expansive bearing to groundwater prospecting, and that the actual investigation has to be done in the field.

The Hyderabad based National Remote Sensing Centre is responsible for remote sensing satellite data acquisition and processing, data dissemination, and aerial remote sensing. In addition to collating data from the satellite imagery, the Center also undertakes aerial photograph surveillance to fill data gaps accruing from satellite imagery and data overlap wherever required. The NRS Center has the capability to collate, and make available the false colored composites in various formats, imaged from IRS 1B, IRS 1C, IRS 1D, IRS P3, IRS P4, IRS P5 and IRS P6.

The Geocoded formats with azimuthal coordinates marked provide a coverage of approximately 725 square kilometers on 1:50000 scale, and are best suited for relevant data overlays, and groundwater exploration planning. Currently NRSC is supplying data from CartoSat 1, 2, 2A & 2B, ResourceSat 1 & 2, OceanSat, TES, IRS 1D and IMS 1 directly to the users. The Centre is also engaged in executing remote sensing application projects in collaboration with the major end users.

CHAPTER III

AQUIFER CHARACTERISTICS AND THEIR SIGNIFICANCE

An aquifer is a geologic formation, group of formations, or part of a formation that contains sufficient saturated permeable material to yield significant quantities of water to springs and wells. Use of the term is restricted to those water-bearing formations capable of yielding water in sufficient quantity to constitute an economic and viable source. Water held in aquifers is known as groundwater. Groundwater is constantly moving, and is seldom stationary. Two main forces drive the movement of groundwater. First, water moves from higher elevations to lower elevation due to the effect of gravity. Second, water moves from areas of higher pressure to areas of lower pressure. Together these two forces make up the driving force behind movement of groundwater, which is identified as the hydraulic head.

Hydraulic gradient:

In fluid mechanics, hydraulic gradient is the most crucial parameter, which governs all flow characteristics of any fluid, both in nature and in all conduit systems. The hydraulic gradient is a vector gradient between two or more hydraulic head measurements over the length of the flow path. It is also called the Darcy slope, since it determines the quantity of a Darcy flux, or volume of water discharging at the end of the slope. Henry Darcy, a French Engineer, in 1856 enunciated the law that governs the flow of liquids. He built an impressive pressurized water distribution system in Dijon, following the failure of attempts to supply adequate fresh water by drilling wells. The system carried water from Rosoir spring 12.7 km away through a covered aqueduct to reservoirs near the city, which then fed into a network of 28,000 meters of pressurized pipes delivering water to much of the city. The system was fully closed and driven by gravity, and thus required no pumps with just sand acting as a filter. He was also involved in many other public works in and around Dijon. Darcy's law is a constitutive equation (empirically derived by Henri Darcy in 1856) that states the amount of groundwater discharging through a given portion of an aquifer is proportional to the cross-sectional area of flow, the hydraulic head gradient, and the hydraulic conductivity. A dimensionless hydraulic gradient can be calculated between two piezometers.

Water moves from one point to another under influence of the difference in head between the two points(fig.3). This difference in hydraulic head, divided by the distance between the points defines Hydraulic gradient. Henry Darcy has experimentally proved that the flow of water, in terms of quantity (m^3 / time) is governed by the hydraulic gradient. This imaginary slope of the water surface (or the potential of the piezometric surface, the level to which water will rise in a series of tubes embedded in the medium) inherently governs the free movement of water, either on the surface, or under the ground. Slope of the piezometric surface is necessarily imaginary, since water does not have a macroscopic flow, but flows through a tortuous path of pores of various dimensions, crevices, joints and fissures of soil, or other porous, semi-porous medium. The resulting streamlines are considered as laminar, and uniform.

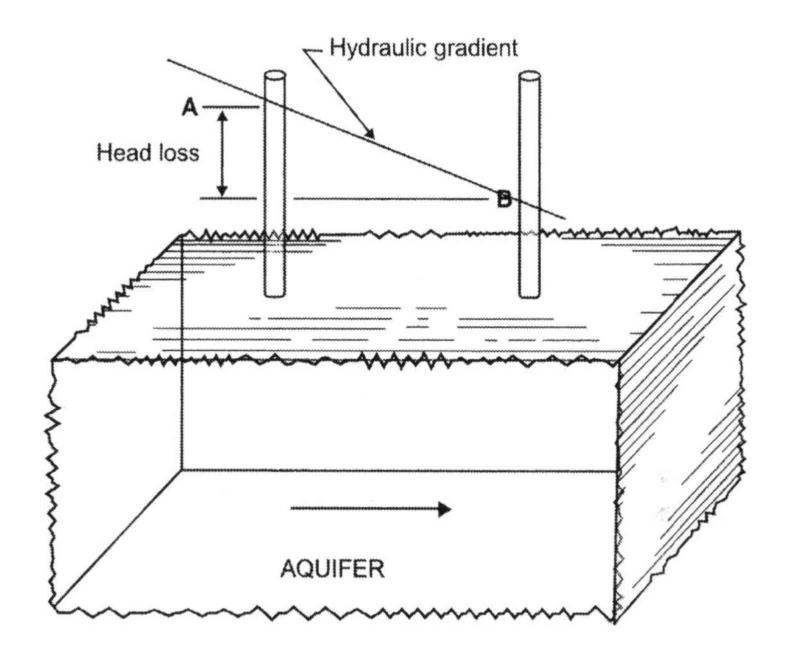

Fig 3: Concept of hydraulic gradient

Hydraulic gradient will vary from place to place depending on the porosity, and permeability of the medium through which the water is migrating. Difference in head, causing the hydraulic gradient is a pre-requisite for water to flow through any medium. Rate of such flow is directly proportional to the gradient resulting from the loss of head between points A and B, divided by the distance between the points.

$$\text{Hydraulic gradient } i = \frac{\text{Difference in head (A-B)}}{\text{Slope length}}$$

Hydrological characteristics of aquifers can be evolved into parameters which, when applied to appropriate formulae, reflect the water yielding capacity of that particular formation. Significant hydrological characteristics are Porosity, Permeability, Transmissibility and Specific yield.

Porosity:

The porosity of a formation is that part of its volume that consists of openings, or pores. In a water bearing formation, porosity indicates how much water is stored in the saturated material.

Porosity in soft rocks is known as primary porosity, inherent to the formation. It depends on the grain size and the arrangement of the grains. Hard rocks develop secondary porosity resulting from tectonic activity, and consist of fractures, joints, and similar lineaments. Porosity is expressed as a percentage of the total volume of the material, or, as a ratio of volume of interstices to the total volume of the formation. For example, if one cu meter of a sandy formation contains 0.25 cu meters of pores, then its porosity is said to be 25 percent. Porosity, while indicating how much water the formation is holding (or will hold under saturation) does not indicate how much of this water the formation will yield, or part with. The formation will yield only a part of the water it is holding, because of molecular attraction and osmotic pressure.

Permeability:

Permeability is a parameter to evaluate the ease with which water will flow through a formation. It is defined as the capacity of a porous medium to transmit water. Water moves from one point to another under influence of the difference in Head between the two points. Henry Darcy has experimentally proved that the flow of water through a column of saturated sand is directly proportional to the difference in hydraulic head between the two points, and is inversely proportional to the length of the distance between these two points.

Based on Darcy's Equation for flow of water, permeability can be expressed as,

$$Q = k \times i \times A$$

Where:

Q = Discharge in a given unit of time
k = Coefficient of permeability
i = Hydraulic gradient
A = Cross sectional area of the formation through which the flow occurs

'k', the coefficient of permeability is the discharge through unit cross-sectional area at unit hydraulic gradient

In soft rocks, the size grading of particles, and the density of material represented by porosity affect flow of water; while in hard rocks the size, the character of the surfaces of the joints, crevices, fractures, and solution openings, called fracture porosity, affect the flow. These characteristics, designated as 'hydraulic conductivity' inherent in any medium is denoted by 'k'; 'k' will change with any variation in any of the features which affect smooth flow.

Permeability can be expressed in cm/sec, m/day, ft/min., or ft/year. Various Units used to express permeability are DARCY and MEINZER Units. They can be converted one into the other by the following factors:

1 darcy = 0.831 m/day
1 cms/sec = 864.0 m/day
1 in/day = 0.0254 m/day
1 ft/day = 0.3048 m/day
1 meinzer = 0.0408 m/day

Determination of permeability:

Permeability of a formation can be determined either in the field, or in the laboratory, by Constant Head, and variable Head methods, which are described in the textbooks on Hydrology and Soil Mechanics. (See references at the end of the book).

Mechanical grain analyses (see chapter on 'Tube wells' for method of determination) also provide data to evaluate approximate permeability values of a loose formation, though there is no accurate way to determine permeability directly from the sand analysis curve. Never the less, with experience, it is possible to estimate the relative water yielding capability of different sands, and sand-gravel mixtures by careful study, and interpretation of the 'effective size', 'uniformity coefficients', and curvature of the analysis curve. Uniformity coefficient is the ratio of the grain size of the 60% of the material to grain size of 10% of the material by weight. The resulting ratio is a measure of the degree of uniformity in the aquifer under analysis. Uniformity coefficient = grain size of 60% material / grain size of 10% of material.

'Effective size' is that particle size, (in mm) where 10 percent of the sand is finer and 90 percent is coarser. Generally, effective size is denoted as D10. Table No.6 indicates the approximate relationship between grain size, and permeability and porosity. Data is abstracted from tests carried out on various sand samples, and can be applied with reasonable accuracy to estimate porosity and permeability.

Table 6: Estimation of porosity and permeability from sand analysis

Effective Size		Uniformity Coefficient	Permeability Coefficient		Porosity Percentage
cms	inches		LPD/ sq. meter	GPD/ sq. foot	
0.02032	0.008	1.2	6700	545	37
0.01524	0.006	1.7	2000	160	35
0.0127	0.005	2.0	1240	100	31
0.02032	0.008	1.6	6460	520	31
0.0381	0.015	3.4	16146	1300	30
0.08636	0.034	1.2	119232	9600	37
0.12192	0.048	1.4	161460	13000	35

Transmissivity:

The expression 'transmissivity' represents the inherent attribute of the aquifer to transmit water, while the term 'transmissibility' denotes the volume of water that is actually being transmitted by the aquifer; though both the expressions have been used one in place of the other. Transmissibility coefficient is the product of permeability coefficient and the thickness of the aquifer. It is one of the important hydraulic characteristics of water bearing formations, which are transmitting groundwater. It indicates how much water is moving through the aquifer. It is conceptualized as the rate of flow of water, in cu.m. per day through a unit vertical section of an aquifer whose height is the thickness of the aquifer itself, and whose width of flow is 1 m, under a hydraulic gradient of 1 in 1 or 100 percent (fig.4). It is expressed as m^3/day/m, or as m^2/day, or gallons/day/ft. Transmissivity can be determined from drill logs and grain size analysis of aquifer samples in the laboratory, or by aquifer performance tests in the field. With this value, it is possible to approximate the volume of water the wells drilled in this formation would yield; following pumping tests on the well, it is possible to estimate and project the drawdown for various pumping rates, over extended time, and the probable interference between wells drilled at different distances from the main pumping well. Pumping tests also provide data, in general, which are representative of a larger area than the single point observations. (See chapter on 'testing wells for their yield and drawdown')

When the coefficient of transmissibility 'T' is inserted into the Darcy equation, the flow through any vertical section of an aquifer is expressed by 'Q' in the equation:

$$Q = T \times i \times w$$

Where:

Q = Discharge or flow in given time (liters per minute)
T = Coefficient of transmissivity
i = Hydraulic gradient
w = Width of the vertical section

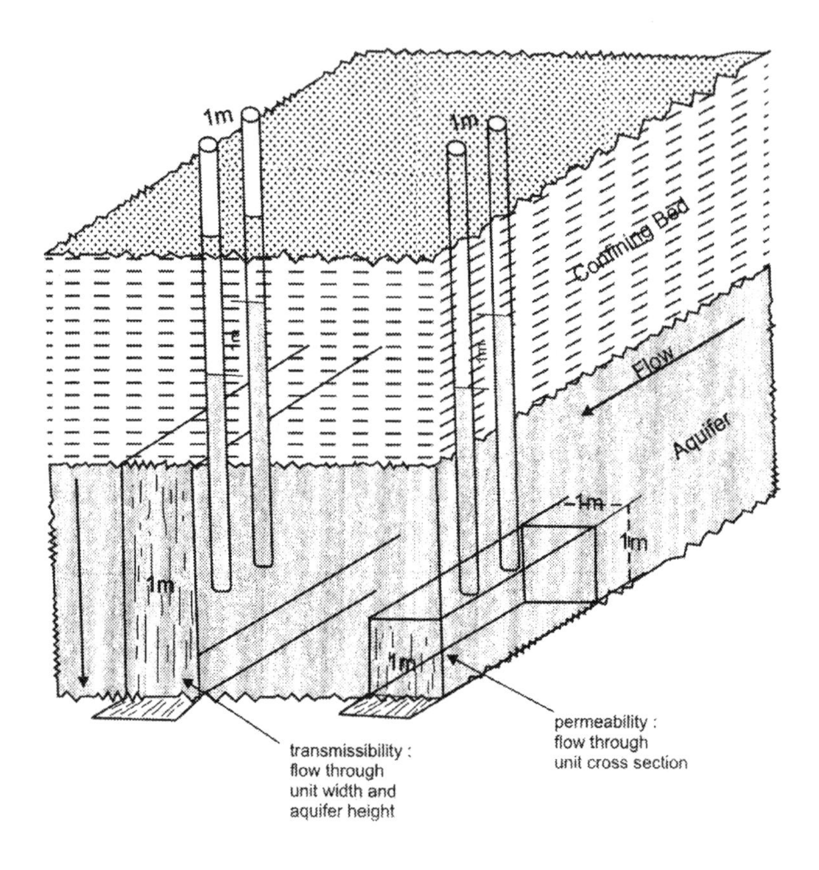

Fig 4: Difference between transmissibility and permeability coefficients

Determination of Transmissivity:

Transmissivity is an abstract aquifer characteristic which can be estimated from aquifer performance tests, or grain size analyses. Aquifer performance test or pumping test involves drilling of non-pumping observation wells in addition to the main well. Long duration tests entail pumping the main well for a period of about 72 hours and monitoring yield and drawdown in this well and drawdown in the observation wells. Recovery is measured in all the wells, and data is plotted on semi-logarithmic graphs to obtain drawdown and recovery curves. Drawdown or recovery during a log cycle, inserted into appropriate formulae provide Transmissivity and storage coefficient values of the aquifer.

Transmissivity values can also be computed, if during drilling drill hole samples have been systematically collected and well logs prepared. A grain size analysis of each of the samples is particularly critical, since the grain size of the aquifer, size variation, along with its silt and clay contamination governs the ease with which water moves in an aquifer. Initial estimates of permeability values are made from this relationship either from the table or a graph drawn plotting 'permeability' against 'effective size' of the sample (Table 6). From the permeability assessment, transmissivity factor is computed by multiplying permeability with formation thickness, obtained from the well log. This geological approach is of significance, where a large groundwater basin is under investigation. A comparative study of the values obtained from the aquifer tests, if conducted, and that obtained from grain analyses, and adding or subtracting the percentage difference will facilitate estimation of 'T' values of production wells without resorting to a series of aquifer tests. This method, to estimate permeability and transmissivity is also pertinent, where conducting the highly structured aquifer test may not be readily feasible.

Data from aquifer performance tests can be projected to obtain, in addition to 'T' values, storage coefficient, boundary conditions, period of safe pumping and safe yield. Never the less, aquifer tests suffer from certain limitations, important of which are: i) field tests are expensive and time consuming; ii) data interpreted is subjective iii) inference has a limited reliability, as it is presumed that the aquifer being tested has a uniform spread without boundary conditions, is isotropic and homogeneous, and is not being recharged, which is rarely the case in nature. The yield and drawdown data obtained either by an aquifer performance test or a single well test is necessary for selection of appropriate pump for the well.

Specific yield and storage coefficient:

Specific yield is a measure of the water yielding capability of the aquifer. It is defined as such in case of an unconfined aquifer, and as storage coefficient in case of a confined aquifer. When water is drained by gravity from a saturated formation, only a part of the water stored within its pores is released, while a considerable part is retained within the formation. This quantity of water expressed in percentage, or in decimals, that a unit volume of the material will yield, is defined as its specific yield. That part of the water retained by the formation is called specific retention. Specific yield added to specific retention is porosity.

As an example, if 0.05 cu m (50 liters) of water is obtained from a saturated formation, its specific yield is 0.05 or 5 percent. If the porosity of the formation is 25 percent its specific retention is 0.20 or 20 percent. Specific capacity and specific retention are dimensionless quantities. One cu m of an aquifer, with a specific yield of 0.10 or 10 percent would yield or part with only 10 percent (10 cu.cms or 100 liters) of the water held in its pores.

Safe yield is the rate at which groundwater could be drawn from a particular formation without producing undesirable effects, such as the well yielding progressively less water. Quantity of groundwater in a given area, and its yield sustenance potential can be estimated from the above factors i.e., volume × specific yield. Seasonal fluctuation in water levels is directly proportional to the replenishment or withdrawal from the aquifer. A systematic course of water level measurements carried out over a given period would provide data to project optimum potential of the area to sustain a given output. Area investigated, multiplied by seasonal fluctuation in water levels or specific yield enables estimation of quantity of groundwater available under unconfined conditions. Excessive withdrawal of water, over and beyond the annual replenishment would lead to incipient depletion of the water table itself, progressively leading to overdraft, and eventual cessation of yield. This is particularly true for hard rock areas. Hence, it is necessary to determine the 'safe yield' factor of the specified area before planning the number of wells.

Specific capacity:

Specific capacity represents the capacity of a well to yield water; it is expressed in liters per minute for each meter drawdown. It partly signifies the competence of the aquifer to yield water to the well and includes the well losses. Specific capacity is extrapolatable data which can be projected to infer aquifer potential in adjacent areas.

Well Hydraulics:

Well hydraulics deal with the water flow systems into a well. Flow of water into a well follows a complex and abstract system, which varies with the nature of the formations through which the flow is taking place. Geological condition of the aquifer through which the water is flowing tends to vary both horizontally and tangentially.

In any groundwater table, whose levels are being lowered because of pumping, water from the surrounding area moves in to replenish the evacuated water. Profile of the sloping water table flowing into, and towards the pumping well assume incipient slopes that gradually become steeper as they approach the point of pumping. This underground flow is a 3-dimensional phenomenon converging from all directions towards the well under pumping. It is a presumptive cone, and is called the 'cone of depression' that takes shape all around the well. Water levels in the surrounding non-pumping wells, if any, which may be reacting to pumping in the main well, reflect a profile of this cone of depression. As the pumping progresses, the depression cone enlarges until the recharge into the well balances the volume of water that is being pumped out. As recharge, and the discharge from the well equal each other, the well attains a state of equilibrium, and the water levels stop declining further. However, in any given geological situation where the cone of depression during its travel approaches a rock boundary, or similar such dry zone, the recharge into the well at that point of time diminishes relative to well's output, resulting in water levels drawing down rapidly. A similar situation arises if there are other wells in the vicinity, which are also being pumped, and cones of depression of any these wells overlap and interfere with each other's passage.

A corollary of the Darcy's law is that velocity of flow varies directly with the hydraulic gradient. When a well is being pumped, the hydraulic gradient created between the pumping well, and the surrounding water table continuously changes as the water levels decline. The hydraulic gradient becomes steeper as the water levels approach the well, and hence the velocity of flow increases proportional to the gradient. It can readily be seen that the profile, the slope, and the behavior of the cone of depression is dependent on the characteristic of the aquifer through which the flow is taking place, and the rate and period of pumping which together contribute to the yield from the well.

Hydraulic conductivity:

Darcy's law essentially states that 'Q' is directly proportional to hydraulic gradient, which in turn depends on the characteristics of the medium through which water is transmitting. Hydraulic conductivity, the factor 'k' is the intrinsic property of the soil or rock to transmit water. It defines the ease with which water can move through that medium. It is in many ways synonymous with permeability, the subtle difference being, while permeability is made use of to compute the quantity and velocity of water that is flowing through the medium based on 'k' values, the expression 'hydraulic conductivity' is connoted to abstract the hydrological property of the medium. In recent years, the term Hydraulic Conductivity is being widely used in preference to Coefficient of Permeability.

CHAPTER IV

WATER WELL DRILLING

There are several methods of drilling for water wells. Drilling methods and procedures vary depending on many factors. Important are:

1. Nature of the geological terrain, physiography and accessibility.
2. Groundwater exploration data.
3. Approximate quantum of water required.
4. Maximum depth of the proposed well.

Amongst the major procedures available to obtain large yields are DTH (Down-The-Hole) drilling and percussion drilling in hard rock regions; and Rotary drilling in soft rock areas.

DTH Drilling:

DTH or Down-The-Hole drilling is the most commonly adopted method to drill water wells in hard rock. It is the fastest, most economical method of drilling, either for an exploratory or for a production well. Since these machines are truck-mounted and self-equipped, they can reach the site swiftly and drill a 300 m to 450 m hole within 24 hours. However, this method is appropriate only for drilling in hard, weathered and fractured formations and is especially inadequate for drilling through soils or soft rock. DTH drilled holes are self-supporting and do not need any casing for a major length of the depth drilled. They intercept fracture and other openings through which groundwater are permeating and tap them for abstraction.

The drill bit is attached to the lower end of a drill string, which consists of a series of hollow 'pipes' with thick strong alloy walls, called drill rods, each of average 6 m length.

Main power source for the operation of the DTH machines is compressed air, supplied from mobile compressors that form an integral part of the drilling unit.

The drill rods carry compressed air to the drill bit at its end. The drill bit consists of an alloy steel hammer, with embedded heavy duty tungsten-carbide cutting edges, which progress down-the-hole by hammering, and breaking the rock into fine chips by both stress percussion, and rotary motions.

Tungsten carbide is one of the most highly abrasive resistant alloys with a potential to cut through very hard formations with little wear. The bits as they drill, and become blunt with use, can also be sharpened by grinding, giving them a fresh lease of life. Most commonly used drilling bits have a diameter of 155 mm, though 200 mm bits are deployed to drill through loose horizons, until hard rock is encountered. The loose horizons are cased to preclude collapse. Drill bits with small tungsten-carbide buttons set in a steel body offer a longer service life. This type of button bit does not require frequent sharpening, and permits more 'down-the-hole' time spent in drilling. However, this type of drilling bit is not economical for drilling in very hard, massive (free of joints and fractures) formations, as its cutting action is subversive to percussion, and hence progress slows to a low. This button bit is particularly valuable for drilling through fractured and heterogeneous formations. In geological terrains where abnormally thick overburden overlies hard formations or where soft rocks superimpose over hard formations, DTH-cum-rotary combination drilling units are deployed.

The contemporary range of DTH drilling machines is capable of drilling 114 to 165 mm (4.5" to 6.5") diameter holes to depths of 300 to 450 meters (1000 to 1500 ft) within 24 hours. Drilling agencies claim availability of machines capable of drilling special purpose holes in hard rock, up to 2 meters in diameter and down to considerable depths (See Ref.).

Some of the additional equipment installed on current water well drilling rigs includes:

1. An injection pump, to force a foaming agent into the hole while drilling in dry formations, to lubricate and cool the drilling bit.
2. A flush pump, to wash the hole and keep it clear of cuttings, and to minimize the formation of a mud cake along the well walls.

Casing depth:

The well extends through overburden, into fresh rock at depth. The top loose horizons in soil or overburden are cased with blank steel, or PVC casing of suitable strength to prevent the soil from collapsing. The lower parts of the drill-hole, drilled through hard rock, or consolidated zones (viz., weathered, and disintegrated rock called murum) are left open. The top zones are cased to prevent unconsolidated formations from sloughing into the well and to check contaminants and surface water from accessing the well. The casing should extend at least 50 cms above the ground level and the space between the outside of the casing and inside of the hole should be tightly sealed. The weathered and underlying hard formations are self-supporting and should not be cased. The blank casing should not generally extend below an average depth of 10 meters below ground level. If the loose soil horizons extend to depths greater than 10m, then it is advisable to install slotted PVC casing below 10 m, to the depth where weathered or fresh rock level is reached. This precaution of not casing below 10 m depth or installing slotted casing below 10 m is essential to tap the water transmitting through these top zones. However, precaution should be ensured to seal the top 10m with blank casing, to ensure that water from the vadose zone, which may be contaminated, does not find its way into the well.

Mechanics of drilling:

Primary power for the drilling operations comes from the air compressor. Compressed air performs three major functions:

1. Operates the pneumatic hammer at the end of the drill string causing both a percussive and rotational action for the drill bit.
2. Evacuates the rock cuttings accruing at the bottom of the hole.
3. Evacuates water, as and when the hole encounters a moisture-bearing horizon.

Drilling pressure:

The drilling bits have specific diameters to suit the diameters of the respective boreholes. They are also synchronous with the air pressures to be deployed for drilling. The diameter, the depth and the speed of drilling is dependent on the pressure and volume of air that the compressor is capable of delivering through the drill-string to the hammer, and the volume of air that can be exhausted to evacuate the drill cuttings to the surface. Reciprocally, the compressor pressure to be deployed is determined based on the nature and hardness of the formations to be drilled. Compressors used with the majority of the DTH rigs have air delivery capacities ranging from 6 m^3/minute to 10 m^3/minute, with point air pressures ranging from 7 kg/cm^2 to 12 kg/cm^2. Air pressure is sustained to rotate the bit at an average speed of 20 to 40 RPM, with the percussion rate ranging between 10 and 20, while drilling in hard, abrasive rocks. Any short fall in the pressure, and volume of air supplied to the bit, as a result of compressor malfunction, or choking of the hammer air exhaust vents will immediately affect the drilling progress.

In the beginning, and at shallower depths, volume of air supply is about 6 m^3/minute (210 cfm) to drill 165 mm (6.5 inches) holes, and 3 m^3/minute (110 ft^3/minute) for 115 mm (4.5 inches) holes. As the drilling progresses, to evacuate rock cuttings effectively from the bottom of the drill hole, as they are formed, the velocity of air exiting the drilling bit, and traveling upwards between the drill string, and the well walls should exceed 900 m/minute (3000 ft/minute).

Typically, as the drilling advances, compressed air is injected at normal pressures ranging between 6 and 7 kg/cm^2 with pressure increasing with depth to about 8 and 9 kg/cm^2. Drilling at higher pressures results in rapid penetration and sensational progress, but causes harmful after effects, such as damage to the drill bit and the hammer, and inclining the hole off the true vertical. High pressure drilling is also essentially counter-productive as, in addition to reducing the life span of the hammer and the bit, excessive air pressure prevents groundwater from accessing the well, providing a misleading image of a dry or low yielding well.

High residual air pressures exerted on the well walls exceed the atmospheric pressure by many times. This force tends to drive the water in the rock openings back through the crevices to considerable distances away from the well. Compressed air tends to remain within the well and openings in the formation for considerable time, preventing the groundwater from accessing the well. Hence, superseding the rate of drilling, drilling pressures should be limited to 8 or 9kg/cm^2. Sometimes, groundwater occurs under confined pressure in an artesian environment, when high pressure drilling becomes indispensable.

As drilling advances to deeper depths, moisture encountered by the hole, and mixed with powdered rock, and rock cuttings results in the formation of sludge, which while moving to the surface, along the annular space between the drill stem and the well walls, forms a filter cake that plasters the well walls and seals the fracture openings. This incipient but gradually coagulating rigid lining forms by a filtration action, when air pressure draws water out of the sludge and squeezes it into the openings in the rock, converting permeable fracture porosity into an impermeable formation. This sludge cake lasts between 12 and 72 hours after completion of drilling. After this period, it usually disintegrates as the water in the fractures breaks into the hole. This phenomenon can be noticed, particularly where high pressure drilling has been resorted to, and where drill holes failed to yield any water during drilling. Wetting of the drill cuttings being ejected is indicative of presence of water bearing formations. Fast, high pressure drilling would seal the water, within the zone and drill through the horizon, with insufficient time for the water to exit. However, incidence of water in the hole would be apparent some hours after cessation of drilling. Hence, in what may appear as a dry hole during drilling, it is always advisable to check for occurrence of water 72 hours after completion of drilling. If water is apparent in the well, its level should be monitored for 3 or 4 consecutive days before deciding to abandon the well.

Major advantage of DTH drilling is that the driller is able to observe, and estimate the depth and quantity of water that is exiting the hole at any point of time. While noting the depth, volume of water discharging from the hole should be measured and recorded by placing a V-notch across the overflow channel.

Water yield, and its depth of occurrence recorded during drilling, facilitates selecting a test pump, and the pump's placement elevation, for a yield test; or if a test is not to be conducted, this data (yield during drilling and the depth where water was encountered) helps purchase, and installation of the appropriate pumping system for the well. Most of the wells yield 40 to 50 percent more water during pumping, than its air displaced water output during drilling.

Groundwater does not generally occur in hard igneous rocks such as granites (exception being volcanic formations such as Traps) at depths greater than 250 to 300 meters; 'striking' water at deeper depths is not entirely due to its occurrence at these depths. It is more due to water from higher levels reaching these drilled depths, and being ejected out by the decreased drilling air pressure. As detailed earlier, very high drilling air pressures tend to seal groundwater inside the 'fracture porosity' in the rock. As the drilling sludge dries and breaks, and the ambient air pressure inside the well declines, this sealed water breaks its way out. Developing a well by pumping air (see chapter on well development) into the freshly drilled well is important. This 'time consuming' but critical procedure called flushing, facilitates in tapping the full potential of the well.

Hydraulic rotary drilling:

Hydraulic rotary drilling is deployed to drill through soft, semi-consolidated, and unconsolidated formations. Penetration is achieved by rotary motion of the drill bit, by trailer-mounted drilling machines, which generally have the capacity to drill 15 to 45 cm (6" to 18") diameter holes to depths of 1000 meters. Oil drilling machines, drilling by the same method, can drill down to great depths, both on-shore and offshore.

Dependent on the cohesiveness of the top formations the initial diameter of the well is drilled with a large diameter 'fish-tail' drilling bit. Further down, usually at a depth between 30 and 50 meters, the well diameter is reduced, and the bit changed to a fast turning rotary drill bit.

The drill bit is attached to a heavy 2 to 3 meter long drill collar, which forms the lower end of a drill string. The drill string consists of a series of hollow, thick walled 'heavy pipes' called drill rods, each of average 6 m length. The drill collar, with its diameter larger than the drill string connects the drill rods and the drill bit. It exerts perpendicular pressure on the bit and provides vertical stability for the drill hole. Drill collars also function as stabilizers, which stabilizes the hole from veering off vertical.

The drill rods are made of solid high tensile alloy and withstand high compressive and torque pressures. Drilling fluid, or drilling 'mud' is pumped under pressure through the drill pipes, which jets through the rotary bit's outlet vanes, and carries the drill cuttings upward in suspension through the annular space between the drill stem and the hole at velocities which range between 0.7 and 1.0 meter/second. On reaching the surface the fluid is channeled through a settling tank to a pit, from which it is pumped back, and re-circulated into the drill hole.

Two types of drill bits are in use: the 'Roller-Bit' and the 'Fish-tail Bit' or the 'Drag-Bit'. Drag bits are employed for drilling in loose, unconsolidated formations, while the Roller Bits are used to drill through the soft rocks such as sandstones and shales. Rotation speeds range between 30 and 60 RPM depending on the hardness of the formations.

Drilling fluid consists of soft clays or special clays such as bentonite mixed in water to different viscosities. Density of the mud bearing fluid is maintained between 1.02 and 1.14 gm/cc depending upon the looseness of the formations being encountered by the drill string. The drilling fluid while circulating through the hole, serves the following functions:

1. Forms a filter cake on the well walls, and prevents it from sloughing into the hole.
2. Flushes cuttings from the bottom of the hole; cools and cleans the drill bit; lubricates drill-bit bearings and the mud pump.

Holes drilled by rotary methods are not self-supporting. They tend to collapse eventually, unless appropriately cased.

CHAPTER V

DUG WELLS, BORE WELLS, RADIAL WELLS

It is possible to source groundwater through two major systems. Foremost is the well known method of obtaining groundwater through a water well. Water wells are either excavated to construct a dug well, or drilled to construct a bore or a tube well. Generally, wells drilled in hard rock, by percussion are called a bore well, while wells drilled in soft rocks with rotary method is called a tube well. An open well draws water from shallow phreatic aquifers, while a tube well taps water from deep lying confined and unconfined aquifers. More complicated are structures that strictly do not fall within the purview of groundwater, but draw underground seepage water from areas adjacent to major water sources. These are the infiltration galleries, intake structures and collector or radial wells, which draw water concurrently, both from the surface and underground sources. These structures are also outlined here, because of the important role they play in obtaining water.

A water well is an Engineering structure needing certain basic parameters for its design and construction. Design of a well has as its objective a combination of optimum yield, long service life, and an economic cost of construction. It is not good technology to construct a well capable of yielding 750 LPM (10,000 GPH), when the requirement is only 250 LPM (3300 GPH). A high yielding well, whose productivity potential is under-utilized, only encumbers the owners with higher initial and recurring pumping and maintenance costs. Conversely, heavy-duty wells supplying water to a Settlement, Industry, or for Irrigation should be carefully designed to yield the maximum quantity of water, with high efficiency in terms of specific capacity. Such wells should also necessarily ensure trouble free, uninterrupted supply, since an unforeseen disruption in water yield may mean reduced production with cascading problems.

Dug wells:

Dug wells are the most simple and economical method of sourcing limited quantities of groundwater, despite their construction being time consuming, and labor intensive. Since the excavation is open cast, many unidentified factors may be encountered during its execution. However, a large number of the unknown factors could be reduced to a minimum by the presence of a Geologist, or in his absence, by gathering hydro-geological data from other similar wells in the vicinity. Dug wells have the unique advantage of being amenable to visual inspection providing easy data regarding quantity of water available at any point of time, time required for recuperation, and seasonal level fluctuations.

Dug wells are feasible only in physiographic and geomorphologic lows, regionally peneplained areas, where rocks have undergone deep weathering, fracturing, or disintegration resulting in thick soil formation. The grading, ill-defined zones between soil and underlying rock are usually rich sources of groundwater. These zones define the depth of dug wells.

These wells, unless excavated in hard murum, need steining, which may be of masonry, brick or concrete. Irrespective of the construction material, steining below the water table should be provided with weep holes which penetrate into the surrounding formations to a reasonable distance. Weep holes near to the entry portal should be packed with a porous filter and screened with a fine mesh. Weep holes and similar other openings afford easy ingress of groundwater into the well. A circular well for the same cross section exposes less area for seepage than a square or rectangular well, but a circular well is economical and safer to construct as a circular section retards caving, and consequently needs thinner steining. In stable areas, square or rectangular well should be preferred. It is sound practice to provide ledges at 3m depth intervals. Thickness of steining increases with depth. Table 7 specifies, for average top soil conditions, the mandatory thickness of steining at various depths for open wells of 2.4 to 6.0 m diameter.

Table 7: Well steining thickness (after Raghunath H.M.)

Depth below ground level	Steining (cms)	
	Brick work	Stone masonry
down to 3 m	33	30
3 to 9 m	46	38
10 to 12 m	53	46
13 to 15 m	61	54

Advantages of Dug Wells:

1. Cheaper to construct; does not need costly equipment and skilled personnel.
2. The structure has a certain capacity to store water.
3. Can be pumped by installing an electric or diesel centrifugal pump on the surface, or at different elevations to match high and low water levels.
4. Water yielding zones are generally visible, enabling occasional cleaning.
5. Wells can be revitalized by blasting, removing debris or drilling at its base or its sides. Instrumentation and maintenance schedules are not called for.

Disadvantages of dug wells:

1. Is a great inconvenience where space available is restricted.
2. The construction is slow, labor oriented, and at times hazardous.
3. Drought seasons affect the water levels immediately, making a dug well suspect, and an unreliable source.
4. Subject to seasonal water level fluctuations.
5. Increase in cost of construction with depth.
6. Deep lying water yielding zones cannot be tapped.
7. Method of construction, if not technically supervised, may seal off shallow aquifers, restricting water flow into the well.

8. Is not dependable for requirement of large volumes of water.
9. Susceptible to contamination by seeping polluted surface waters.
10. Cannot be reliably tested to predict yield and project safe pumping limits.

Dug wells are the main stay of rural economies all over the world. However for various reasons have not been bestowed the importance they deserve.

Bore wells:

Wells drilled in hard rocks are frequently called 'bore wells'. These are narrow vertical wells of 115 mm or 165 mm (4.5 or 6.5 inches) diameter, drilled in hard rock to various depths. Top soil or unconsolidated zones are blank cased, to depths generally not exceeding 7 meters, to prevent caving in of top soil. The casing depth is limited to the minimum required, as any excess casing will seal the top water yielding zones, and prevent sub-surface water from reaching the well. The hole below the cased depth is left exposed to allow direct inflow of groundwater into it. A properly calibrated multi-stage submersible pump is deployed to pump water from deep wells, while jet pumps are used to pump water from wells of shallow and medium depths. Output from these wells depend to a large extent on the method of drilling adopted, correct selection of the appropriate pump, and the cumulative yield from the fracture porosity intercepted by the well.

A dependable and sustained water supply from a bore well is conditional on the following:

1. The well point is located on a proper groundwater source
2. Situation is free of any potential surface contaminating sources
3. Appropriate pump is installed in the well.

Correct location implies, apart from being drilled in the proximity of an investigated source,the well point is so located as to preclude contamination from adjacent surface sources. The well should preferably be situated higher than the surrounding ground surface, to ensure good drainage. Placing the well mouth above the surrounding surface could be achieved by placing a concrete slab of about 50 cm thickness around the top casing. Exit piping is also placed within the concrete slab. All possible sources of contamination should be at a lower elevation than the well, and the distances to such potential contaminating sources should not be less than 5 meters, and not less than 30 meters from septic tanks. Bacterial contamination of a water supply generally occurs when seepage from sewage systems or surface water accesses the well. Contamination may enter the well through the top or by seeping along the well walls. Tests have shown that bacterial contamination is usually eliminated after water has filtered through 6 meters of average top soil. The well is constructed with the top 6 meters of casing made watertight.

After completion of drilling to the targeted depth, wells need adequate development to ensure that they yield optimum volume of water. After completion of sufficient development, the wells may be depended on to yield optimal quantities of water over several years irrespective of the rainfall.

Advantages of bore wells:

1. Requires less space.
2. Can be constructed quickly.
3. Reliable yield of large quantities of water.
4. Can be tested to select and install the correct pump; possible to predict its safe yield and future behavioral characteristics. If properly maintained, water levels are not affected either by seasons or drought years.
5. Once drilled, properly developed, and the pumping systems are properly calibrated, the well can be depended upon to provide uncomplicated service for several years.

Disadvantages of bore wells:

1. Requires mechanical equipment to construct.
2. The Drilling contractor should be reliable and experienced. This single essential rationale can ultimately result in the life, success, or failure of a well.
3. The well should be technically sophisticated; requires owner's supervision during drilling.

4. Wells drilled with worn bits; and drilled off plumb will not permit pump assembly to be installed at the correct depth.
5. Requires expensive pumping and piping equipment.

Design of a bore well:

Following points need consideration while planning a bore well.

1. Availability of space.
2. Approximately estimated quantum of water required.
3. Data on seasonal fluctuations of water levels in the area.
4. An estimate on the cost of construction.

Analyses of the above factors will broadly indicate the type and dimensions of the well to be constructed.

Well diameter:

Cost of construction multiplies exponentially with well diameter. Hence, particular attention should be paid to the cost aspects, though an increase in diameter, results only in a marginal increase in the yield from the well. However, an enhancement in diameter facilitates installation of a higher capacity pump to draw more water. The diameter should be large enough to accommodate the pump selected for the requisite yield and head, with a minimum 3cm clearance between the bowl assembly and the inner wall of the well. A slightly larger diameter facilitates lowering of the pump assembly, particularly when the well is off plumb.

Effect of well diameter on yield:

Increase in well diameter has no significant influence on the yield, either from a bore well or a dug well. Field and mathematical analyses (Dupuit's equation, 1863) for steady state flow conditions prove that the yield of a well does not depend as much on the diameter of a well, as it does on the radius of its influence on the enveloping water table. The Dupuits assumption holds that groundwater moves horizontally in an unconfined aquifer, and that the groundwater discharge is proportional to the saturated aquifer thickness.

Table 8 indicates the marginal increase in yield for various well diameters, when radius of influence=120 m.

Table 8: Volume increase in yield with enlarged well diameters (after Johnson)

	Well Diameter (cms)						
	15	**30**	**45**	**60**	**75**	**90**	**120**
Yield in LPM	500	550	585	610	635	655	685
		500	530	555	580	595	625
			500	520	540	560	585
				500	520	535	560
					500	515	540
						500	525

1. It can be seen from Table 8, that if a 15 cm diameter well yields 500 LPM, a 30 cm well constructed at the same situation will yield 550 LPM; and a 45 cm well yield 585 LPM.
2. Doubling the diameter of a well will augment its yield by about 10 percent.
3. To double the yield from a well, diameter needs to be increased by about 60 times. It could be concluded that there occurs an insignificant increase in yield with any enlargement of the well diameter.

Well depth:

Depth of any well is a critical factor, which is decided based on the required yield, and the hydro-geological conditions of the area. Knowledge of the local geological features and pumping data from the currently functioning bore wells nearby is an important pre-requisite prior to correlating depth and probable yield of a well. Resistivity survey data, the drillers log, and the data from the wells in the vicinity usually indicate the number, thickness, and the yield potential of aquifers that needs to be penetrated to yield the required output. Maximum depth of bore wells is generally limited to 300 meters. Further down, volume of inflows into the well decreases rapidly with depth. However, in basaltic terrain, no such depth restraint is applicable as groundwater inflows into the well swell directly in consonance with the number, and thickness of vesicular and amygdaloidal flows, and the inter-trappean horizons the hole intercepts. Basaltic flow thicknesses, including massive flows, which are devoid of any porosity, seldom exceed 10 meters. Hence, in basaltic terrain, it is advisable to drill as deep as possible to tap as many inter-trappean and vesicular flows as feasible.

Radial/Collector Wells:

These wells, also known as Ranney wells after its inventor, are constructed in permeable alluvial formations, or close to surface water sources such as rivers and lakes to tap seepage from the surface. It abstracts water from an aquifer with direct relation to a surface water source like a river or lake. The amount of water available from the collector depends more on the surface water source than to the piezometric surface of the aquifer, though these wells typically draw water both from the surface and subsurface sources. Many major rivers and large streams, during monsoon floods silt up their beds, and during the dry season flow through their own silt deposits. Collector wells are designed to draw water flowing through such silt and sand beds, and along the flood banks of rivers. The water drawn from a radial well represents abstraction from the surface waters, which are typically infiltrating into the ground, as also the groundwater from the shallow water table that subsists proximate to surface water source. Appropriately located and designed, such wells have the capacity to provide abundant supplies of water for large-scale exploitation such as for a water supply scheme, Industry or a large farm. However, it is important that the location and feasibility of such systems be technically well investigated. By conducting an aquifer performance test on a test well drilled on the banks of the source stream, it is possible to compute the transmissibility of the shallow aquifer, and project the dependable yield from a Collector well, at diverse locations with varied effective radii. The flow system around a radial well is very complex, and the simplest approach is to treat the system as flow to a vertical well with a radius that is about 75 percent of the lateral extent of the Collectors (Mikels and Klaen, 1956).

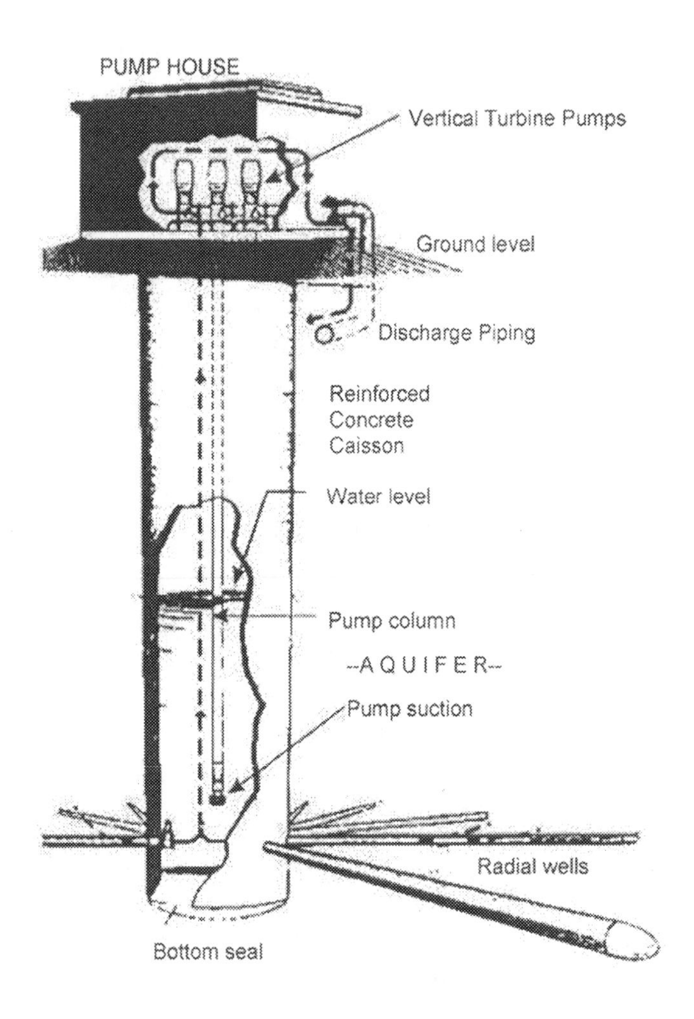

Fig 5: Collector and radial wells (Mrinalini et al)

Collector wells consist essentially of a large diameter (4 to 7 m average) shallow well of reinforced concrete. Wall thickness averages 45 cms. From this caisson, 10 mm thick slotted and screened steel pipes extend radially into the surrounding formations to collect water, and lead it into the central well from which it is discharged with heavy-duty pumps (fig.5). Collector wells built along riverbanks will have their radials extending hemi-spherically only from the circumference facing the river to tap water flowing below the riverbed, while similar wells installed in the river bed will have lateral strainers extending radially all round the well. Such radial extensions have been computed to enhance the effective diameter of the well, by as much as 75% of the length of the laterals. Pumping from collector wells induces seepage from the surface water source, and will in course of time evolve into a reliable source, delivering large quantities of both seepage and surface waters.

The lateral pipes, which extend out radially into the surrounding sands, range from 15 to 50 cms in diameter. They are perforated or slotted to acquire about 20% open area to allow easy inflow of water through the strainers. They are terminally fitted with rigid steel driving guides which help to force them through the surrounding sands by special hydraulic jacks, which can drive them to distances exceeding 30 meters. This method of constructing these wells is the reason for calling them Jack wells. Short spans can be hammered to lengths of about 15m. Successive lengths of the pipes are butt-welded until the planned length is attained. These laterals can be placed at different levels based on the aquifer incidence, and the output required. After placement, the pipes are washed and flushed until a natural filter develops around them, and ingress of material from the surrounding aquifer ceases.

Collector wells for coastal areas:

In deltaic aquifers along coastal areas, fresh water being less dense floats over the denser saline water. The higher water levels inland cause the freshwater to constantly flow seaward. Coastal aquifers exposed at depth discharge this fresh water into the sea by virtue of the hydraulic gradient sustained by the sources emptying into the sea. At the sea-land boundary, freshwater flows out from the highest point of the aquifer, and at the lowest point, saltwater flows into the aquifer. Seawater seeping inland (beneath the fresh water) forms a mixed interface with the fresh water. This boundary between the lighter fresh water and the denser seawater slopes down into the land, while the fresh water lens formed by water infiltrating into the coastal aquifer gently slopes towards the sea. This results in an uneven wedge of fresh water bordering the seacoast.

Seawater has a specific gravity, which varies from 1.025 to 1.04. Because of this variation in specific gravities of fresh and salt water, the floating wedge of fresh water maintains a hydrodynamic balance with the underlying seawater. Generally, this balance causes fresh water to extend to a depth of average 40 times the thickness of fresh water above the mean sea level. This means that for every 1 m or 1 ft thickness of fresh water above the sea level, there occurs a 40 m or 40 ft thickness of fresh water floating over the saline seawater. This relationship is of critical significance and was discovered independently by a Dutch scientist Baden-Ghyben and a German scientist Herzberg, and is known as the Ghyben-Herzberg principle. This relationship is of great importance, since water supply to populated areas along the coast and on islands, depends largely on this hydrostatic balance between the sea and fresh water. The underground boundary that separates the fresh water layer from the salt water is not a sharp boundary line. In reality, this boundary is a transition zone of brackish water (fresh water-salt-water mixture), caused by seasonal fluctuations in rainfall, tidal action, and the amount of fresh water being withdrawn.

The flow of fresh water and continuance of the hydraulic gradient is a pre-requisite to maintain the fresh water-seawater interface at some depth below the land surface. Further inland, the freshwater column is higher due to the increasing altitude of the land, and the higher column equalizes the pressure from the salt water, preventing saltwater migration into land. Pumping of fresh water inland reduces the water pressure, drawing salt water into new areas. The salt water being mineral rich and dense, pushes inland under pressure; and as freshwater levels drop, saltwater intrusion proceeds further inland.

For every 1m depth of fresh water that is dewatered, above sea level from a coastal aquifer, the saline water interface rises 12 m. Consequently, the indefinite boundary between the two is in a constant mode of flux, because of reasons, such as rainfall, water withdrawal and tidal surge.

Consequent to infiltration of fresh water into the ground, either from rainfall or from a surface water source, the boundary depresses into the ground resulting in a significant increase in thickness of fresh water floating above the saline water. As fresh water enters the water table, it concomitantly moves saline water more towards the sea. Along the Saurashtra and Kutch coasts, restricting withdrawal of floating fresh water from wells, and allowing the rainwater to replenish and raise the water table has been one of the most effective methods in containing the sea water from ingressing inland.

It has also been seen that this occurrence of 12m thickness of a fresh water lens over the interface may be true for isotropic confined aquifers; but is not exactly valid for aquifers, which deviate from the normal. There is no frontier for variations in nature. The incidence of occurrence of 40 units thickness of fresh water for every 1 unit of fresh water above the sea level is however the guiding principle for exploration for groundwater along the coast and up to considerable distances inland.

Wells drilled along the coast, either for exploration or for production, as they penetrate the transition zone, cause severe salt-water intrusion, and contamination of the fresh water aquifer. Hence, drilling a well along the coast is fraught with hazard, unless extreme care is taken to drill short of the transition zone. However, in areas which have been well investigated, and the depth to the transition zone is well known, bore wells drilled short and sealed well above the boundary, and screened in the upper reaches, yield significant quantities of fresh water. However, output should be carefully controlled by pumping with a low capacity pump. Any turbulence at the seawater—fresh water interface will cause the underlying saline water to break into the overlying fresh water aquifer, causing contamination of the water being drawn. Detailed monitoring of the pumping, and study of lab models is undertaken in many countries to assess how much water near to a coast can be safely pumped without triggering sea water intrusion.

Radial wells are particularly well suited for coastal areas, where fresh water, occurring only above a certain level is potable and clean. Shallow collector wells, also known as skimmer wells, 4 to 6 m in depth are constructed to tap fresh water flowing in low-lying coastal aquifers in deltaic areas. The radial strainers of appropriate length skim and filter the water draining through the river deposits and divert them into the caisson for centralized pumping. Radial collectors placed only in upper fresh water zones prevent salt-water incursion into the fresh water aquifers. A gravel filter and/or a nylon mesh over the strainer slots are essential to preclude ingression of sand and silt. Monitored drilling, and pumping tests to explore the thickness and extent of the aquifer in the delta, and to determine the saline water—fresh water interface is an essential pre-requisite for design of the system. Saline water-fresh water interface comprises a fluctuating, abstract boundary, dissimilar from a defined contour.

Meager water supply to Lakshadweep islands is through Collector wells, located where fresh water lenses are at their maximum. As a sequel to pumping the shallow aquifers, the saline—fresh water interface rises during pumping, and declines during recovery, necessitating regulated pumping periods. Water supply schemes in Malta and many other islands resort to such radial wells for town water supply.

Chennai is a prominent metropolis on the east coast of India, which has been consistently facing the problem of insufficiency in its water requirements over several years. Recent studies, concluded over an investigative period of ten years has indicated a rapidly rising sea water-fresh water interface which will inexorably lead to a further deterioration in the City's worsening water situation. This is a result of a large part of the population depending on groundwater for their daily needs, in the absence of suitable infrastructure for regular water supply. Further studies may indicate that the answer to the city's water problems may lie in strictly controlling over pumping of the numerous wells tapping into the meager fresh water lens that floats on sea water.

Infiltration galleries:

Infiltration galleries are extended radial wells of considerable length. They do not tap groundwater in the strict sense of the term, but draw water from the surface sources seeping underground. They are constructed immediately adjacent to a river, or directly below the river bed itself, either parallel or across the flow to collect and conduct significant flow of water occurring below the riverbed, to a pumping caisson. Infiltration galleries consist of large diameter, perforated concrete pipes with a gravel filter shroud. It is laid in an excavated trench, with the trench bed gently sloping towards the central caisson, a graded filter placed around it, and the trench backfilled. Inspection manholes with hatches are essential along the length of the infiltration gallery.

CHAPTER VI

TUBE WELLS

Drilling and construction of wells in soft formations are more complicated than those in hard rocks. These wells called tube wells are also deep, constructed in loose and soft formations, and are designed to draw groundwater in large quantities ranging up to 2000 liters per minute. Tube wells in soft rocks are complex, expensive structures, involving detailed investigations, planning, intricate testing, design, and specialized machinery, and the presence of experienced drilling Engineers and site Geologists. Such complex wells become necessary, if the geological environment is made of 'soft' sedimentary formations and requirement of volume of water is large. Depending on its dimensions, and the hydro-geological conditions, well-constructed tube-wells have the potential to provide vast quantities of water for agriculture, town water supply, and industrial application. Tube wells are cased throughout their drilled depth. Based on the geological log and the stability of the formations, the diameter of the well is reduced with depth. Reciprocally casing diameters also decrease and telescope one into the other with depth. The casing consists of blanks interleaved with suitably designed slotted screens placed against aquifers. Surrounding the screens are gravel packs to filter and prevent aquifer material sloughing into the well.

Construction:

Construction of a tube well needs heavy drilling Equipment, well logging instruments, and test pumps. Drilling and post-drilling formation and borehole logging is an integral part of the construction procedure. Procedure involves:

1. Geological and geophysical logging to determine hydrological characteristics of the aquifer; caliper logging to determine the profile of the drilled hole, radiometric logging to detect presence of radon gas in case of town water supply projects, and careful collection of drill-hole samples for laboratory analysis.
2. Well design to decide depths at which casing diameters should decrease, and telescope into the next lower casing
3. Preliminary yield test (PYT) to determine yield characteristics of the well
4. Aquifer performance test to determine transmissivity, specific yield, and storage coefficient
5. Selection of pump for the production well
6. Study and selection of aquifers which could be screened to draw the optimum yield
7. Seal aquifers with unpotable water, or water with radicals unsuitable for irrigation, or for any particular industrial application
8. Analysis of aquifer grain size to optimize screen dimensions to control entrance velocities
9. Calibration of gravel pack size, aquifer grain size, and screen openings
10. Placement of gravel shrouds around the production assembly, or in case of light duty wells, insertion of pre-gravel packed screened casing.

Well casing and screens:

All wells drilled in recent and soft, unconsolidated, and semi-consolidated formations need to be cased, through their entire effective depth, to retain the hole from collapsing. Screened casing is set against water bearing sands, while the rest of the formations are blank cased. The well screen is a specialized piece of equipment consisting of closely spaced openings, which permit easy ingress of water into the well. The screens are made with openings of various sizes and shapes to suit the grain size and gradation of the water bearing formation. The screen, because of its design, prevents fine, loose material from the aquifer finding its way into the well. The proper selection and installation of the screen largely determine the efficiency of the completed well.

Determination of the grain size of the aquifer sands is a pre-requisite for screen size selection, which is obtained from mechanical analysis of the samples, systematically collected during drilling.

The representative samples are geologically classified, and mechanically analyzed to select the aquifers to be screened, from which water can be drawn into the well.

Mechanical analysis:

Screen slot size, and the grain size of the gravel packs shrouding the screens are selected based on the size composition of the aquifer. Mechanical analysis, also called sieve analysis is an important procedure to determine this composition.

The grain size distribution in an aquifer sample is found by passing a representative sample of the aquifer through a set of sieves, and weighing the quantity of fractions retained on each sieve. A set consists of a series of sieves starting from the one with the coarsest opening, and successively decreasing to a sieve with the finest possible mesh. The sieve openings are standardized by ISI (IS:460-1962), with their respective openings, and designations in millimeters. IS sieve sizes range from 4.75 mm to 0.075 mm.

British standard ASTM sieves, designated in inches, (ASTM E 11-1961), from 2 to 200 are also in use.

Normal procedure is to select about six sieves to suit the sand sample. If the sample contains gravel, the coarsest sieve to be used is the IS 4.75 mm (equivalent to ASTM 4), below which sieves with smaller openings are placed to grade the sample into fractions of smaller grain size. A typical set would have IS 4.75mm, 2.00mm, 1.00mm, 0.425mm(425 μ) 0.212mm, 0.125mm, and 0.075mm(75 μ).

An alternative method of mechanical analysis of sands employs determining settling velocities of the grains, based on the fact that coarse grains settle faster in water than the finer grains. A mechanical analysis of a sample of sediment can be made by allowing the material to settle through a long column of water and observing the amounts of material that reach the bottom of the column during successive experimentally determined time intervals. A complete analysis requires about five minutes for most sands.

The sample for sieve analysis would consist of at least 400 gms of oven dried soil devoid of segregations, or lumps. A weighed quantity is placed in the top coarse sieve, and the entire set is run through a mechanical shaker for about 12 minutes. The quantity of sample retained in each of the sieves is weighed to arrive at the following:

1. Percent retained on any sieve = (Weight of soil retained) / (Total soil weight × 100)
2. Cumulative weight retained on any sieve = sum of weights retained
3. Cumulative percentage retained on any sieve = sum of percentages retained on successive sieves
4. Percentage finer than any sieve size = 100 percent minus cumulative percent retained

Table 9: Sieve analysis of an aquifer sample

Original Weight = 400 gms		
Sieve opening (mm)	**Cumulative weight retained (gms)**	**Cumulative percent retained**
1.18	60	15%
1.00	84	21%
0.60	200	50%
0.50	244	61%
0.425	292	73%
0.355	336	84%
0.212	380	95%
Pan	396	100%

Table 9a: Plot data for sand analysis curve (fig.6)

Grain size (mm)	0.21	0.35	0.42	0.50	0.60	1.00	1.18	**'X' axis**
Percent retained	95	84	73	61	50	21	15	**'Y' axis**

This data can be plotted in various ways to establish respective parameters for screen size selection. Cumulative percentage retained on each sieve is plotted on a graph paper against the respective sieve openings to obtain a curve (fig.6).With practice, it is possible to predict, from the trajectory of the curve the grain size variation in the sample. Analysis data can be plotted either on logarithmic or arithmetic graphs, but profile of a curve with relation to grain size variation is better identified on arithmetic graphs.

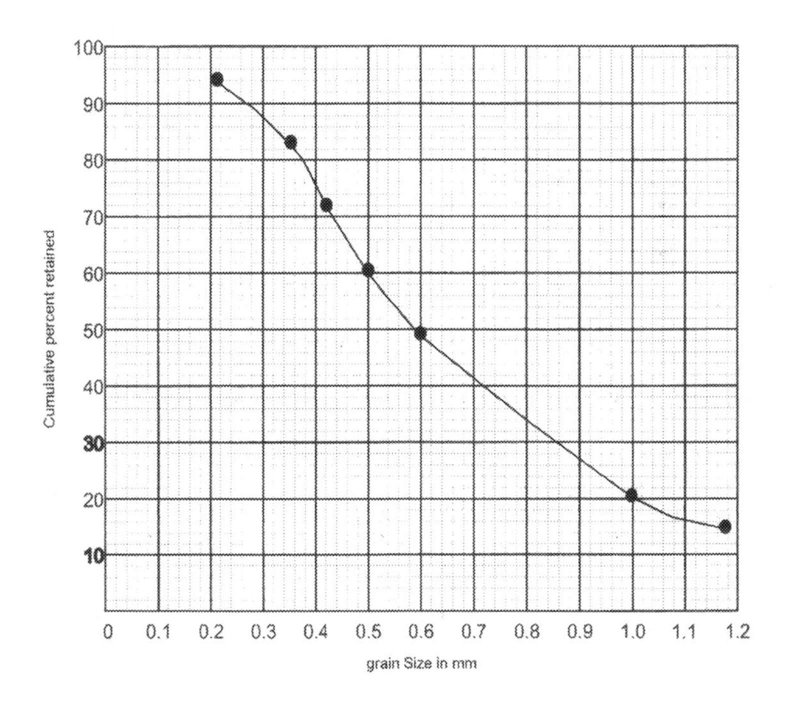

Fig 6: Grain size analysis curve

Effective size:

The curve in fig.6 represents grain size analysis of an aquifer sample. The graph shows that 90% of the sand consists of grains larger than 0.25 mm, and 10% is smaller. This specific point in the curve, expressed as a grain size, above which the grain sizes are finer, and below which the grains are coarser, is the effective size, a term developed by Allen Hazen. Effective size defines the fineness or coarseness of the sand, and is made use of to correlate the aquifer with its permeability, and its water yielding potential. (See chapter on permeability)

Uniformity Coefficient:

Allen Hazen developed the concept of Uniformity Coefficient to identify the average slope of the major length of the curve, between the 90% and 40% particle sizes of the sand sample. It is defined, and calculated as the quotient of the 40% size of the sand divided by its effective size. Lower its value, uniform is the grading of the sand between these limits. Large values denote less uniform grading. In the sample analyzed (fig.6) 40% grain size (0.75 mm) divided by its effective size (0.25) is equal to 3, which is its Uniformity coefficient.

Sand analysis curves can be plotted in many different ways, based on its application. Another method is to plot the percent passing various sieves. To plot such a curve the percent retained is subtracted from 100. Any aquifer horizon consists of a wide range of particle sizes from the largest to smallest, with intermediate sizes distributed in varied ways. Hence to obtain a picture of the grading in any aquifer, three parameters need to be integrated: 1. fineness, 2. slope of the curve, and 3. shape of the curve. Each of these factors is independent of the other. Sand analysis curves assume many different shapes depending on any combination of these factors.

Fig 6 is a typical grain analysis curve integrated from the mechanical analysis of a sand sample from an aquifer. Below, in a tabular format is the result of the sand analysis:

The curve in fig 6 is constructed from the cumulative percentage of sand grains of various sizes retained on different screens as illustrated in Tables 9A and 9B.

Selection of well screens:(excerpted from Edward E.Johnson Inc., Groundwater and wells Johnson screens Division)

Efficiency of a well system depends largely on the casing and screen design. Appropriately calibrated slots filter and allow sand free water to flow into the well in optimum quantities, without causing excessive loss of Head (fig.7)

A typical screen consists of:

1. Openings in the form of slots, which are continuous around the circumference of the screen. Slots are essentially inverted V-shaped, which widen inwards.
2. Close spacing of slot openings to provide maximum percent of open area consistent with adequate strength of the screen to withstand soil pressure.
3. Single metal construction to prevent galvanic corrosion.

a. well screen against ungraded sand

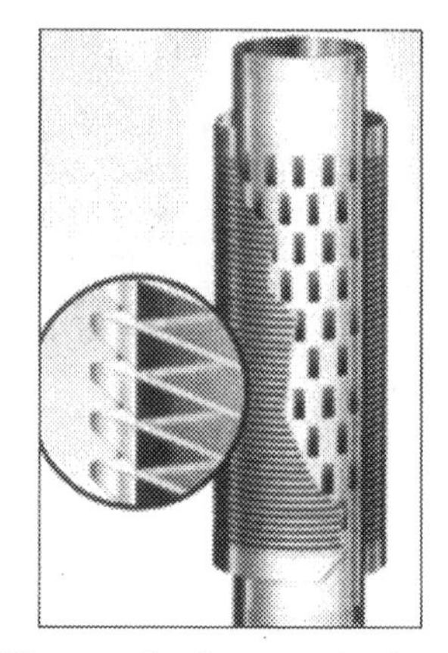

c. Wire wrapped well screen against fine sand.

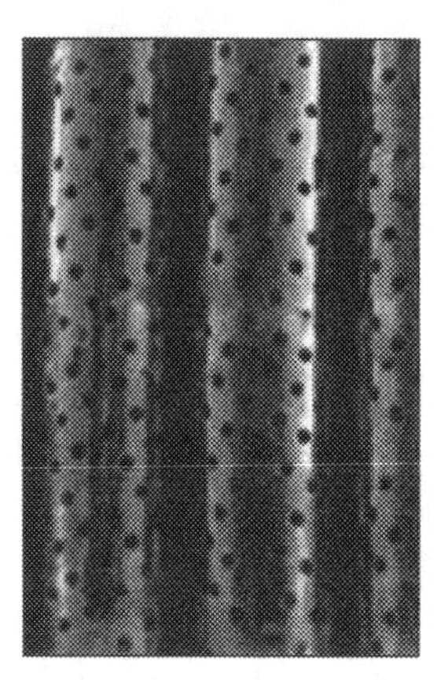

b. screen against coarse sand and gravel

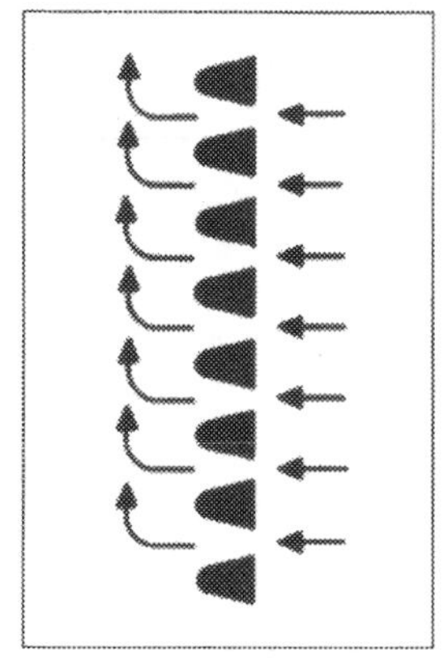

d. Enlarged screen section showing non-clogging cone shaped openings.

Fig 7: Typical well screens

Fig 7 depicts some models of screens, available for diverse types of aquifers with varying grain sizes. Fig 7 'a' shows a universal screen for unsorted sands. Fig 7 'b' is of a random screen, with rounded openings for use against coarse sands, gravel and sheared rock formations.

Fig 7 'c' illustrates slotted screen, wrapped with wire windings for use against fine and medium grained sand and sheared or fractured rock formations. Enlargement gives an idea about water jetting through the wire winding. Fig 7 'd' illustrates enlarged section of a normal screen with non-clogging V-shaped openings and controlled water intake that will not allow aquifer sands to migrate into the well. Screen selection should be based on the aquifer grain size analysis.

In a well under continuous pumping, water flows more freely through a screen with a large open area. The entrance velocity of water through a large intake area is low, and consequently the ensuing head loss is minimum. This arrangement effectively reduces the drawdown in the well for a given rate of pumping.

Slot opening:

The dimensions of the slot opening are determined based on the grain size, and its gradation in the aquifer. To establish these openings, a grain analysis curve is plotted for each of the aquifers intercepted by the drill hole. In naturally developed wells, the slots are kept open by 40 to 60 percent of the average grain size of the aquifer sands.

To determine the appropriate slot opening, a point is selected on the graph, where the 40 percent (or 60 percent) axis intersects the analysis curve, and from this point, move down the vertical axis to reach the corresponding grain size. Slot openings should be of this size, which would retain an opening between 40 and 60 percent of the sand.

For the sample analyzed, the slot opening would be 0.72 mm if the screen is to retain 40 percent of the sand grain size, and 0.50 mm if it is to retain 60 percent of the sand grain size. Higher the quantity of sand to be retained smaller will be the opening resulting in less inflow into the well. A fine balance is made between the quantity of sand that could be allowed to migrate into the well, and the amount of water that could be pumped from the well. The 40 percent size is preferred when the groundwater is not corrosive.

Screen length:

The optimum length of the screen for any particular well is chosen based on the thickness of the aquifer and the pumping water level. Aquifer thickness is determined from the formation log made during drilling, often supplemented by data from an electric log. An electric log brings out the exact boundary between the formations, nature of the formation, water potential and the quality of water.

If the aquifer is homogeneous and confined, the optimum length of the screen is based on the following general principles:

1. If the aquifer is less than 10m thick, screening length is limited to 70% of its total thickness.
2. If the aquifer thickness varies between 10 and 15m, then 75% of its total thickness is screened.
3. If the aquifer thickness exceeds 15m, then 80% of its total thickness could be screened.

Screen lengths corresponding to the above guidelines tap 90% or more of the specific capacity that could be tapped by screening the entire aquifer.

The screen lengths are further governed by the effective open area of the screen, and entrance velocities. Equation 5 inter-relates the various factors, and is of value while deciding screen lengths:

$$L = \frac{Q \times n \times 6.5}{A_e \times V_c}$$

Where:

L = Length of screen in meters
Q = Discharge in liters per minute
A_e = Effective open area per meter length of screen (approximately one half actual open area)
V_c = Critical velocity in meters per minute (velocity above which sand particles are moved into the well in case of naturally developed wells)
n = Aquifer thickness screened.

In the case of non-homogeneous confined aquifers, it is best to screen the most permeable strata.

The hydraulic conductivity of the aquifer is correlated to soil properties like pore and particle size (grain size) distribution, and soil texture. In a case where all samples have about the same effective size, the curves with the steepest slopes frequently specify the most permeable material. A steep curve represents more uniformly graded sand. Greater uniformity of grain size enhances permeability, other factors being identical. In the field, it is even possible to visually study the samples obtained during drilling, and estimate their relative permeability values from their coarseness and cleanliness.

Effective open area of a screen is generally one-half the actual open area, but varies with the screen type, and can be obtained from the screen manufacturers who provide Tables showing the open screen area for various slot sizes. The percentage of open area in slotted pipes range from about 1 percent for 0.75 mm slots to about 12 percent in 6 mm slots. Certain designs have 18 percent open area.

Entrance velocity is the velocity with which water enters the well through screens. This velocity is restricted to a critical limit, above which the water will excite and transport the sand from the aquifer into the well. Entrance velocities depend on the permeability of the aquifer, and the open area of the screen; entrance velocities should be limited to 3 to 7.5 cms/sec varying with the open area of the screen. The average velocity varies from quite low at the bottom of the screen, to many times more opposite highly permeable zones in a naturally developed aquifer. Table 10 shows in both meters per minute, and centimeters per second the critical entrance velocities for various permeability factors.

Table 10: Permeability & critical entrance velocities (after Johnson)

Critical velocity (max)		
Permeability(litres per day/m^2)	**meters/minute**	**cms/second**
205000	2.7	4.3
162000	2.4	3.8
122000	2.1	3.4
101000	1.8	2.9
81000	1.5	2.4
60000	1.2	1.9
41000	0.9	1.5
20000	0.6	1.0

Laboratory, and field tests have shown that following advantages accrue whenever the entrance velocity is limited to 3.0 cms/ second, or less:

1. Friction losses resulting from water entering the screen openings becomes negligible.
2. Rate of encrustation reduces to a minimum
3. Rate of corrosion reduces to a minimum.

The entrance velocity is calculated by dividing the expected yield of the well by the total area of openings in the screen. If the figure is greater than 3 cms/sec, the screen diameter is increased to provide sufficient open area so that the entrance velocity is less than 3 cms/sec.

Summarizing, the screen length is determined by the aquifer thickness; the screen openings, and the grain size gradation of the aquifer; while the screen diameter is varied to limit the entrance velocities. Transmitting capacity of a well screen is an integral of the screen itself at a specific entrance velocity, and has nothing to do with the yield from the aquifer.

Manufacturer's screen size selection Tables indicate the nominal screen size in inches or cms, (the size of the steel pipe through which screen may be set by the telescoping method) against intake area per lineal foot of screen in sq.inches, or per lineal 30 cm lengths of screen in sq.cms.Every manufacturer provides similar selections for arriving at the proper screen size to collate with the computed entrance velocity.

Screen Transmitting Capacities:

The transmitting capacity of a well screen is expressed in liters per second for 30 cm length of screen at the recommended entrance velocity of 3 cms/sec. It is calculated by multiplying the sq cms of open area by a factor of 0.003.

For example the open area in a 30 cm well screen with No.40 slot opening is 839 sq.cms. Its transmitting capacity is 839 × .003 = 2.5 liters per second. A meter length of this screen has the capacity to transmit 8 liters per second of water at a velocity of 3 cms/sec. The pump is usually installed well above the intake screens, so that the head loss accruing from vertical flow of water is at its minimum.

Gravel packs:

Most economical well design consists of a naturally developed well, where an aquifer is screened, with the designed slot openings directly against the aquifer. However, under certain situations, discussed below, a gravel pack around the well screens becomes necessary to improve well efficiency. An artificial gravel pack around the screens prevents migration of fine sands into the well. Slotted pipes, and screens with pre-shrouded gravel packs, cemented to the pipes are available for ready lowering into the well. Such artificial gravel shrouded slot pipes are deployed, where natural gravel is not readily available within economical distances, or where time available is short. They are particularly useful, where formations surrounding the well have a tendency to slough around the casing.

A sieve analysis curve indicating slot openings of less than 0.75 mm size implies that despite installing a very fine screen, the aquifer sands will access the well. In such a case, a gravel shroud, as a filter around the screen, becomes an essential requirement. A gravel pack is also necessary, when the aquifer is nearly homogeneous, with a Uniformity coefficient of less than 3, and effective size is less than 0.25 mm.

In a naturally developed well, the zone in the immediate vicinity of the intakes is pumped out to create a more permeable zone, while in an artificially gravel packed well, the fine material is replaced with gravel of appropriate size creating an interface to filter the fine sands. In either case, the net hydraulic effect is equivalent to enlarging the effective diameter of the well with substantial enhancement in water yield. Providing a gravel pack against a fine-grained aquifer has the viable advantage of installing larger slot openings, which would enhance substantially the well efficiency.

Four advantages of an artificial gravel pack are:

1. Amplifies the effective radius of the well.
2. Specific capacity is enhanced as a result of the increased entrance velocity.
3. Sand migration into the well is minimized, particularly in fine, loose formations.
4. Enables provision of a larger slot opening against fine, highly permeable formations, and where chemical quality

of the groundwater is suspect, suggestive of a potential for encrustation.

Loosely cemented, or semi-consolidated formations are also frequently gravel packed to preclude possibility of well walls sloughing onto the screens.

Design of gravel packs:

Gravel packs increase the efficiency and life of a well. They are specifically designed to correlate with each individual aquifer from which the well will be drawing water. Basic parameters for an appropriate design are obtained from sieve analysis curves, which are constructed for all strata comprising the aquifer.

The effective size ratio of gravel pack to formation known as the "Gravel pack ratio" is obtained by.

$$\text{Gravel pack Ratio (GpR)} = \frac{50\%\ \text{gravel pack}}{50\%\ \text{aquifer}}$$

The permeability of the gravel pack should generally be between 3 and 5 times the aquifer permeability; Wells with the gravel pack ratio between 4 and 5, generally have a high output efficiency; while, wells with GpR between 7 to 10 are less efficient; and when GpR exceeds 10, wells tend to pump considerable amounts of sand. The feeder formation need not possess a low uniformity coefficient to be effective, when packed with uniform sized gravels.

Gravel size selection:

Design procedure to determine proper gravel size selection is based on the grain size analysis of the aquifer material. Gravel size is obtained by multiplying the 70% size of the aquifer material by factors given in Table 11. The product of this multiplication is plotted as the 70% size of the gravel on an arithmetic graph. Through this initial point is plotted a smooth curve, by trial and error method, to derive a uniformity coefficient of 2.5 or less. The sizes represented by this curve would constitute the range of the gravel pack.

Size of gravel is generally retained at 4 times the grain size of the aquifer, if the formation is fine grained and homogeneous, and 6 times the aquifer grain size, if the formation is coarse and heterogeneous. (Table 11 for marginal variations). The screen should possess sufficient strength to withstand pressures exerted by the gravel. In corrosion free waters, PVC pre-gravel packed screens can also be used. Reciprocally, gravel packs provide additional support to the screens against collapsing aquifers.

Table 11: 70% Size multiplication factors for gravel size selection

Aquifer	Size (mm)	Uniformity Coefficient	Multiplication Factor
Fine grained uniform sand	0.07 to 0.25	2	4
Medium grained ungraded sand	0.25 to 0.50	2-3	5
Coarse grained heterogeneous sand	0.50 to 1.00	5	6

In fig. 6, 0.40 mm is the 70% retention point, which multiplied by 6 indicates (0.40 × 6) 2.40 as the 70 percent size of the gravel. This forms the initial point on the gravel-pack analysis curve. From this point, a smooth curve corresponding to a uniformity coefficient of 2.5 or less is drawn on the graph. The permissible range in grain size variation is indicated by this curve. The gravel pack curves should correspond to as much uniformity in size as is practicable. The percentage of each of the sizes is determined by selecting 4 or 5 sieve sizes that would correspond to the spread indicated by the curve, with about 10% tolerance. This assortment would largely constitute the expansive size distribution of a typical gravel pack. (Edward E.Johnson Inc., Groundwater and wells: Johnson Divn., UOP Inc., Minesota).

Example:

Sieve analysis of an aquifer sample collected during drilling indicated the following critical grain size characteristics:

Effective size = 0.04 mm
40% size = 0.20 mm
Uniformity coefficient = 5
70% size = 0.10 mm.

Retention factor, 70% size of the aquifer is 0.1 mm. Multiplying this size by the factor 6, (Table 11) 0.6 mm is obtained as the 70% size of the gravels that would constitute the gravel pack. This point is plotted on the graph, as the initial point, through which is fitted a smooth curve with an effective size of 0.45 mm, and a 40% retention size of 0.85 mm which yields a uniformity coefficient of 1.90. Material representing, on the average, the various sizes in respective percentages of the material indicated by the gradation curve is the combination that is selected to shroud the screen.

Method involves selection of screen sizes, which encompass the spread of the curve. A variation of 8 to 10%, either below, or above the percent retained on each of the sieves is assumed as within the permissible range.

Thickness of gravel packs:

Laboratory studies made by Edward E.Johnson, Inc., have shown that very thin gravel packs measurable in millimeters successfully restrain the formation sands from entering the well, irrespective of the velocity of water flowing into the well. Hence, the thickness of the gravel pack is mostly directed by the technical viability of placing a thin pack. Average thickness is 7 cms, which however can be limited to 3 cms in case of ready made pre-packed well screens.

In theory, a pack thickness of 2 or 3 grains is all that is required to restrain formation particles from migrating into the well. However, in the field, it has been found that it is necessary to place 7 to 10 cm thickness of gravel pack around the screens to ensure a stable envelope. Upper limit of thickness of the gravel pack is 20 cm; any thickness above 20 cms makes final well development difficult with unsatisfactory results. To provide a gravel thickness exceeding 10 cms, cost of drilling escalates proportionately. Gravel packs with a thickness of less than 5 cm act more as formation stabilizers, working to support the aquifer and are not effective as a filter.

Formation stabilizer:

Drilling for water by rotary method typically involves a drill hole 10 cms larger than the diameter of the slotted casing, which furnishes about 5 cms annular space around the casing. This additional space provides the required clearance for trouble-free installation of the production assembly to the bottom of the drill-hole. This annular space is backfilled by coarse sand to filter fines slumping from the higher horizons during development of the well.

Formation stabilizers consist of sand-gravel mixtures, or clean coarse sands, without complex gradation as in the case of artificial gravel packs. However, they need to be slightly larger than the aquifer grain size. Sands intended for manufacture of mortar is suitable as formation stabilizer for use with a wide range of aquifers. These stabilizers are placed around the casing to a level of about 10 meters above the screens. During development of the well, as the stabilizers migrate into the well considerable slumping will occur which needs to be replaced.

Screens for corrosive waters:

Corrosion of well screens by groundwater is one of the major causes of well failure. To safeguard against such deleterious collapse, the chemical nature of the groundwater should be determined, and screens made of material capable of resisting such action needs to be installed.

Incidence of following chemical contents indicate corrosive nature of the groundwater, portending eventual well failure:

1. Acidic waters with pH value less than 7.0
2. Dissolved oxygen exceeding 2 ppm; common in shallow water wells.
3. Presence of hydrogen sulphide; detected by characteristic rotten-egg odor; less than 1 ppm of hydrogen sulphide can lead to severe corrosion

4. If TDS (total dissolved solids) exceeds 1000 ppm electrical conductivity will set in, causing electrolytic corrosion. This can be obviated by installing well screens made with a single corrosion-resistant metal.
5. Carbon di oxide generated by carbonates in the groundwater, in excess of 50 ppm results in generation of mild carbonic acid leading to corrosion.
6. Presence of total chlorides exceeding 500 ppm.
7. Dissolved oxygen exceeding 2 ppm

Encrusting waters:

Ground waters with the following chemical characteristics, in course of time, tend to deposit an encrustation over the screen openings in cased wells, and over the pump intake filters in uncased wells. Such encrustation clog the entry ports, and constricts flow into the well, leading to progressive well failure.

1. High pH values exceeding 7.5.
2. Waters with carbonate hardness exceeding 300 ppm
3. Iron content exceeding 2 ppm.
4. If manganese content exceeds 1 ppm in alkaline oxygenated waters, manganese precipitates and clogs the screen openings, and pump inlet filters.

Treatment against encrustation:

Mineral precipitates crusting over and clogging the screens and pump filters can be removed by introducing concentrated hydrochloric acid. However, the well screens, and the pump intakes should have been made with metals, which can withstand such acid treatment. Iron bacteria, though by itself is not injurious to health, produce a jelly like, slimy substance, in addition to oxidizing and precipitating dissolved iron and manganese. These slimy substances plug the pores of the aquifer, clog the well screens, and pump intake filters. If iron bacteria occur in the groundwater, they can effectively make the well cease yielding water within a short time. They can be eliminated by use of chlorine to destroy the organisms, followed by acid treatment to dissolve the precipitated iron, and manganese.

CHAPTER VII

WELL DEVELOPMENT, STIMULATION, AND MAINTENANCE

One of the most important operations in completing a well, is the development of the water bearing formation to facilitate the aquifer to yield optimum quantities of water it is holding or transmitting. Development is an integral part of any water well drilling program. Significance of this phase of the project, and its impact on the water yielding sustenance of a well is very important, and merits emphasis.

Well development is carried out after completion of drilling. It consists essentially of removal and elimination of drilled remnants, mud, and other fines from the well itself, as well as from that part of the aquifer, which immediately surrounds the water entry points. Development is carried out by intermittent pumping of air at various depths of the hole.

Development in soft rocks:

In wells drilled in soft rocks, development is carried out after lowering the educator assembly and gravel packing the well. The well is thoroughly washed by pumping water through the drill string. This is achieved by intermittent pumping, using a piping system devoid of a foot valve. The development is achieved by alternately moving water in the well into the aquifer, and back from the aquifer into the well. The most commonly applied technique is 'surging', consisting of moving a heavy bailer or a close fitting surge block up and down the well, making the water to flow into and out of the aquifer. This procedure also facilitates breaking and removal of the drilling mud that would cake the well walls, and obstruct aquifer water from flowing freely into the well. Occasionally, while drilling through soft formations which tend to collapse, it becomes necessary to use high density drilling mud to prevent collapse. In such a case sodium hexametaphosphate, a high potent dispersing agent is introduced into the well during the development process. This chemical breaks the mud cake on the well walls and the mud choking the gravel pack.

Another effective surging technique is air surging, which consists of pumping air into the well under pressures of about 10 kg/cm^2. The development assembly is made of two concentric pipes, one placed within the other. The inner pipe is the air inlet, and the outer pipe serves as the eductor. Air from a compressor is impelled into the center pipe, which forces water up through the annular space between the air pipe and the eductor. The entire assembly is lowered into the well, and moved up and down while blowing compressed air; the air inlet pipe is also independently moved within the eductor until maximum quantity of sand free water is obtained.

Proper development results in the creation of a zone, which has coarse grains proximate to the screens, and gradually grading back to the undisturbed aquifer resulting in the creation of a natural filter around the well.

Development in hard rocks:

In case of wells drilled in hard rocks, after completion of drilling, development of the well is crucial to tap the optimum potential of the well. Groundwater occurs in hard rocks in joints, fractures, crevices, and interfaces between various zones of weathering. Water held in these zones normally occurs under ambient holding pressures exceeding the atmospheric pressure by varying degrees.

The use of compressed air, from the rig's compressor, under proper conditions, is a rapid and effective method to develop the well. Using the drill rods as an air inlet, compressed air is injected at various levels and the water blown out, until it runs clear of sludge and mud. The standard 6 m^3/minute air delivery is adequate to develop medium depth wells. Air pressure should range up to or more than 8 kg/cm^2 which can air lift about 500 lts per minute of water from the well depending upon the pipe submergence.

Development consists essentially of surging, jetting, and similar such action, together called 'flushing' to cause agitation close to the well walls. This results in opening of the fissures, crevices, joints and similar other channels in the rock; and the surging action moves any disintegrated or broken fine material stuck in the fissures and joints into the well, permitting the water to flow freely into the well. This flushed material is pumped out by the compressed air circulating in the well. In addition, the drilling action generates rock powder, which on mixing with water results in a cement like sludge. The exhausting air forces this sludge into the openings in the rock and effectively seals the openings, precluding water from entering the well. This clogging of the water entry channels can be broken up, and cleaned by a sustained development process. It involves two procedures. First is pumping in air under high pressure. After completion of drilling to the projected depth, as each drill rod is withdrawn, air from the rig compressor is pumped in under high pressure for about 10 minutes before withdrawing the next rod. As the rods are being taken out one-by-one, the drill string keeps becoming shorter, and as the end of the drill string rises up in the well, high pressure air exiting at the end of each of the rods will impinge on the well walls. This high pressure air jetting cleans the wall faces and flushes it of the sludge coating; and in the process opens the fractures, and crevices releasing groundwater flow into the well. Second part of the flushing process consists essentially of cleaning the well to bring out all the drilled sludge that would have settled into the well. Reasonably clean water exiting from the well is an indication that adequate flushing has been done.

Running a close fitting rubber packed surge block, up and down the well is another method of developing the well. The surge block helps in clearing the well of sludge blocking the openings. In the absence of a surge block, the well should be flushed to its entire uncased depth by running a water or air jet under pressure through the well, to clear the extraneous material that accumulates during drilling. The optimum output from the well could be obtained only if all the fractures and crevices are prodded and stimulated to contribute their maximum yield.

After flushing the well, and withdrawing the drilling machine, an appropriate pump (see—Pump selection) is installed at the pre-selected depth. Development of the well is continued with the pump, by intermittent pumping for a minimum of at least 3 days, until the water runs free of rock powder.

Well stimulation:

Stimulating a well to yield water to its maximum potential is an important component of the post-drilling operations. There are various processes of improving the yield from a well, vaguely grouped under the heading 'stimulation'. Stimulation becomes meaningful only when:

1. The well yielded deficient quantities of water during drilling, and subsequent yield test.
2. The preliminary exploration for groundwater has given positive indications for occurrence of water.

'Stimulation' broadly means prodding or provoking the well to yield more water. This becomes particularly necessary when the yield during drilling has been unsatisfactory. Some of the well stimulating methods are simple, which could be undertaken by the user himself, while there are other advanced methods, requiring specialists in the field. Amongst the simpler methods are:

1. Further development of the hole by jetting compressed air through the length of the drilled hole, by reinserting the drill rods.
2. By introducing or injecting external water (5000 to 10000 liters depending upon the depth of the well) into the well, allowing it to decline and pumping it out, either by compressed air or a regular pump.
3. It is good practice, in case the borehole failed to yield any water during drilling, to verify the hole, 48 hours after drilling, for presence of any water and if present, its level. If incidence of water is evident, then steps a) and b) above should be undertaken instead of simply abandoning the well. Such 'dry' wells have subsequently yielded large quantities of water in due course of time.

Advanced methods of well stimulation are undertaken when:

1. The requirement of a water source becomes imperative at a particular point, such as in the proximity of an industrial unit.

2. Despite development, output from the well is found to be unsatisfactory.
3. Stimulation becomes necessary when, over a course of time, a gradual decline is observed in the water yield from the well.

There are many stimulation methods, but the most effective and common are the blasting and acidizing techniques, both of which need experienced technicians to handle the procedure.

Explosives set off in the well at carefully selected depths often enhance the yield by enlarging and widening the openings, thereby creating a prone semi-venturi condition.

Acid treatment is indicated when a gradual reduction in yield is noticed in wells:

1. That are cased with screened sections;
2. That are located in calcium bearing formations such as limestones, and dolomites;
3. That are drilled in the vicinity of calcium bearing formations.

In any of these situations, calcareous minerals carried by the groundwater precipitate and grow a crust on the screen and pump entry filters, gradually clogging the portals, resulting in diminishing output. Harder the well is pumped, higher the rate of encrustation.

Concentrated hydrochloric acid with a suitable inhibitor dissolves the crust with ease. The acid is introduced into the well through a pipe. The set-up consists of a 25 mm pipe reaching to the bottom of the well, which is coupled at the surface to a large diameter T with a funnel on top. A wrought iron or plastic pipe is used; galvanized pipe or fittings should not be used. Sufficient acid is introduced, through the funnel into the well to fill about 3 meters height. Then the pipe is raised by about 3 meters, and an equivalent quantity of acid is added, and the sequence continued till the static water level is reached. The acid being heavier displaces the water, but becomes diluted in the process. It is important to maintain acid strength by periodic refilling.

The acid-water mixture in the well is agitated either by an air lift pump, if a rig is available nearby, or by installing the production pump and running it intermittently with the gate valve shut. The quantity of acid piped into the well should be sufficient to fill the entire water column in the well, in case of open wells; and fill the screened sections in cased wells.

Sulfamic acid (not to be mistaken for sulphuric acid), available in dry granular form enabling easy transshipment and handling, can be used instead of hydrochloric acid. Action of this acid is little slower than that of hydrochloric acid, particularly at ordinary temperatures. Hence, it needs to be retained in the well for a longer time. Sulphamic acid is introduced into the well by mixing the granular material with water in a surface based tank, and siphoning it into the well. Some agitation is necessary to dissolve the acid in the water. About 15 kg of granular acid mixed in 100 ltrs of water provides the necessary concentration. Since sulfamic acid has a limited solubility, excess dry acid can be introduced to retain the strength of the solution for a longer time.

Acid treatment also facilitates withdrawal of screened casing, or a submersible pump, stuck inside wells. Acid softens the well walls, and disengages the section from encrusting material that fastens it to the well walls.

Sustaining well yield:

Bore wells and tube wells tend to fail over a course of time, for various reasons. Most important cause is lack of sufficient attention during the construction of the well, viz., inadequate development before commissioning the well. Next significant reason is overdraft (over-pumping), drawing more water from the well, exceeding the natural inflow into the well. This excess withdrawal is enforced on the well by pumping in spurts, which will cause irreparable damage to the system, eventually ending in its total failure.

Monitoring yield from the well:

The well should be monitored frequently, particularly during the change of seasons, to ensure its long life. Most important is to check the yield and monitor the water levels. Any major changes in either or both are usually indicative of an unhealthy well and impending failure. Discharge should be controlled to maintain a desired or particular water level and its corresponding yield.

To monitor these two critical parameters, at the time of construction, the well should be provided with the following permanent fixtures:

1. A pressure gauge and a throttle valve along the discharging arm.
2. A 12mm or ½ inch flexible tubing attached to the educator pipe, ending just above the pump, and inserted into the well along with the pump. The line should be corked at the top to prevent any material or insects falling into it. This tubing, in preference to a water level indicator, will facilitate insertion of a steel tape to check water levels. (see chapter VI)
3. A flow meter in low yielding wells, or an appropriate device to check output in high yielding wells.

This instrumentation on a well is akin to a 'thermometer', which indicates the health of a well.

It is possible to estimate the yield by constructing a 'table' indicating the output pressure against various volumes of flow. With such a table, 'pressure' can be correlated to yield, to facilitate estimation of output directly, irrespective of the closure status of the throttle valve.

CHAPTER VIII

MEASURING FLOW & YIELD

During the groundwater exploration and evaluation projects, it is frequently necessary to measure the quantum of water that a well is discharging. Water flow, yield, pumping rate are inherently synonymous terms to signify water outflow from a well.

Measuring flow:

Flow or water meter:

A commercial type water meter is the most convenient instrument available to monitor flow or pumping rate in permanent installations. The dialed meters show the total volume of water that is flowing through a pipeline in cubic meters or liters. Subtracting two readings at recorded time intervals, say, one minute apart provides the pumping rate. Non-contacting ultrasonic Flow meter from 'Greyline instruments' (www. greyline.com) measures flow from outside of metal or plastic pipes. These instruments are essentially clamp-on ultrasonic transducers, which measure flow and relay the data.

Filling containers of known volume:

A simple method of determining discharge rate is to compute the time required to fill a container of known volume, such as an oil drum or a rectangular container. Following simple equations can be applied to compute the discharge. For high yielding wells, a more complex method of monitoring flow rate, such as manometer or a suitable weir across the outflowchannel is required.

Oil drum of 220 liters capacity:

$$\text{Discharge in liters per minute} = \frac{220 \times 60}{\text{Time in seconds to fill drum}}$$

$$\text{Discharge in gallons per hour} = \frac{48 \times 60 \times 60}{\text{Time in seconds to fill drum}}$$

Rectangular containers/tanks/reservoirs:

$$\text{Discharge in liters per minute} = \frac{\text{Length (m)} \times \text{Width (m)} \times 1000}{\text{Time in minutes to raise water level by 30.5 cms}}$$

$$\text{Discharge in gallons per hour} = \frac{\text{Length (ft)} \times \text{Width (ft)} \times 6.23 \times 60}{\text{Time in minutes to raise water level by 1 foot}}$$

In case of large circular containers/tanks/reservoirs, water flowing into the system can be determined by measuring:

$$\text{Discharge in liters per minute} = \frac{\text{Diameter (m)} \times 1.6}{\text{Time in minutes to raise water level by 10 cms}}$$

$$\text{Discharge in gallons per hour} = \frac{\text{Diameter (ft)} \times 5.26 \times 60}{\text{Time in minutes to raise water level by 1 foot}}$$

OR

$$\text{Discharge in gallons per hour} = \frac{\text{Length (ft)} \times \text{width (ft)} \times \text{rise in water level (inches)} \times 516}{\text{Time in minutes to raise water level by 1 foot}}$$

Measuring flow through triangular and rectilinear weirs (Tables 12 & 13):

Volume of water exiting from wells under drilling can be monitored by channeling the flow through a 90° or 60° triangular notch weir. The channel upstream of the weir should have a section width so that terminal contractions 'a' 'a' is not less than 0.75 l. depth of channel upstream of the weir should not be less than 20 cms below the notch apex. The water in the channel should 'head-up' upstream of the weir, and then overflow through the notch(fig.8).The channel water should not directly overflow the notch. The channel downstream should allow free flow without causing turbulence. Table 12 provides base data of volume of water overflowing triangular notch weirs.

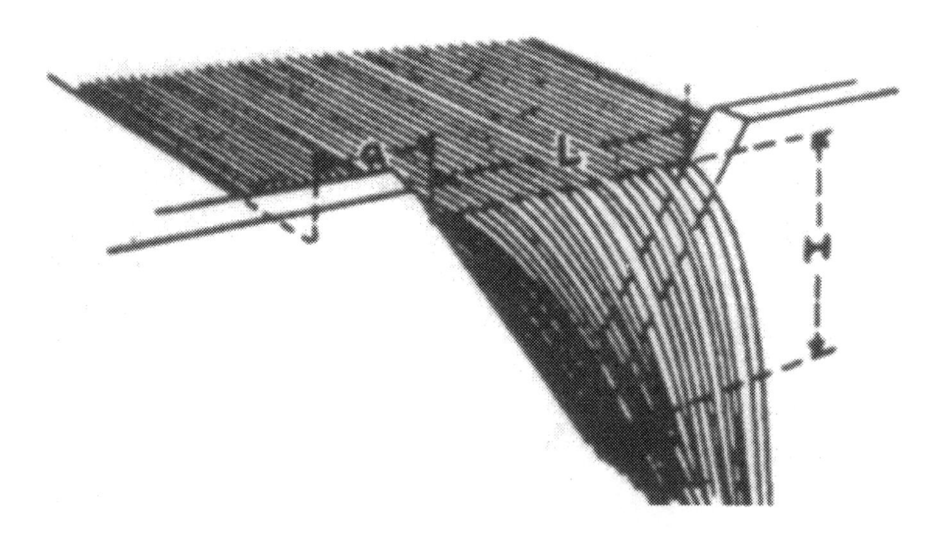

Fig 8: Measuring volume of water overflowing triangular notch weir

Table 12: Volume of water overflowing triangular notch weirs

HEAD OVERFLOWING 'V'		90° NOTCH		60° NOTCH	
Inches	cms	GPH	LPM	GPH	LPM
1¼	3.2	191	14	110	8
1½	3.8	302	23	174	13
1¾	4.4	444	34	256	19
2	5.1	620	47	358	27
2¼	5.7	835	63	481	36
2½	6.4	1085	82	625	47
2¾	7.0	1375	104	795	60
3	7.6	1710	129	985	74
3¼	8.3	2089	158	1205	91
3½	8.9	2514	190	1450	110
3¾	9.5	2984	226	1725	130
4	10.2	3509	265	2024	153
4¼	10.8	4084	309	2359	178
4½	11.4	4714	356	2719	206
4¾	12.1	5399	408	3114	235
5	12.7	6148	465	3539	268
5¼	13.3	6948	525	3999	302
5½	14.0	7798	590	4494	340
5¾	14.6	8698	658	4999	378
6	15.2	9647	729	5598	423
6¼	15.9	10697	809	6198	469
6½	16.5	11797	892	6798	514
6¾	17.1	12997	983	7498	567
7	17.8	14196	1073	8198	618
7¼	18.4	15496	1171	8948	676
7½	19.1	16895	1277	9747	737
7¾	19.7	18345	1387	10597	801
8	20.3	19845	1500	11447	865

HEAD OVERFLOWING 'V'		90° NOTCH		60° NOTCH	
Inches	**cms**	**GPH**	**LPM**	**GPH**	**LPM**
8¼	20.96	21444	1621	12397	937
8½	21.59	23094	1746	13346	1,009
8¾	22.23	24893	1882	14346	1,085
9	22.86	26643	2014	15396	1,164
9¼	23.50	28542	2158	16496	1,247
9½	24.13	30492	2305	17595	1,330
9¾	24.77	32541	2460	18795	1,421
10	25.40	34691	2623	20045	1,515
10½	26.67	39189	2963	22594	1,708
11	27.94	43988	3326	25393	1,920
11½	29.21	49187	3719	28392	2,146
12	30.48	54685	4134	31592	2,388
12½	31.75	60584	4580	34991	2,645
13	33.02	66832	5053	38590	2,917
13½	34.29	73430	5551	42389	3,205
14	35.56	80428	6080	46438	3,511
14½	36.83	87776	6636	50686	3,832
15	38.10	95574	7225	55185	4,172
15½	39.37	103622	7834	59834	4,523
16	40.64	112270	8488	64833	4,901
16½	41.91	121267	9168	70031	5,294
17	43.18	130665	9878	75430	5,702
17½	44.45	140462	10619	81128	6,133
18	45.72	150760	11397	87027	6,579
18½	46.99	161407	12202	93175	7,044
19	48.26	172554	13045	99623	7,532

HEAD OVERFLOWING 'V'		90° NOTCH		60° NOTCH	
Inches	cms	GPH	LPM	GPH	LPM
19½	49.53	184151	13922	106321	8,038
20	50.8	196147	14829	113270	8,563

(computed and edited from base data in 'Water well Handbook by Keith E.Anderson)

For large flows, such as from high yielding wells or stream sections, a rectilinear weir is deployed across the channel. Table13 indicates the volume of water flowing through these weirs, both in imperial gallons per hour (GPH) and liters per minute (GPM) against the respective heights of flow above the weir.The height of the abutting plates on either side of the weir should be not less than 3H, and the length of the weir opening should be at least 4 times 'H' as shown in the figure. Length of weir opening 'L' should be more than 4 times 'H' the height of water overflowing the weir (fig.9)

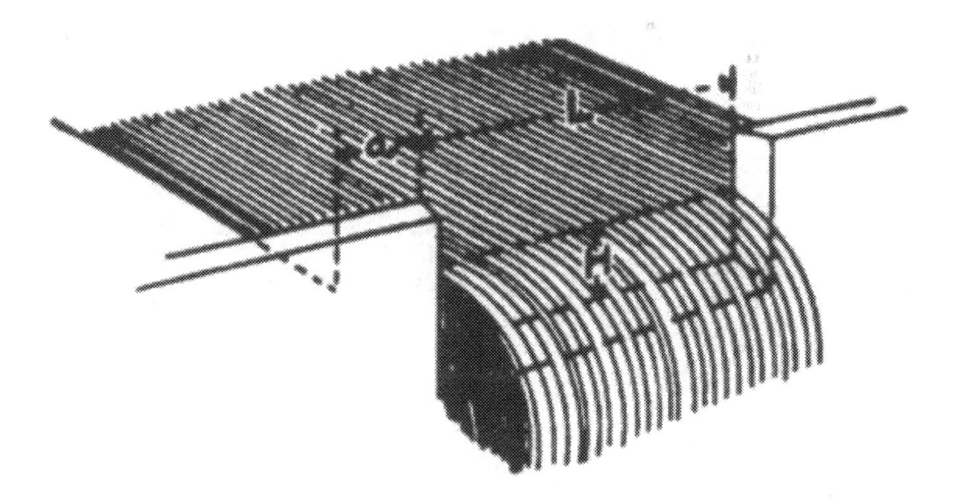

Fig 9: Measuring volume of water flowing through rectilinear weir (courtesy: Keith E. Anderson: water well handbook—original by Ingersoll-rand Company)

Table 13: Volume of water overflowing rectilinear weirs

HEAD		WEIR LENGTH							
Inches	cms	30 cms		90 cms		150 cms		Every 30 cms above 150 cms	
		GPH	LPM	GPH	LPM	GPH	LPM	GPH	LPM
1	2.5	1770	134	5374	406	8988	679	1802	136
1¼	3.1	2474	187	7518	568	12517	946	2519	190
1½	3.8	3244	245	9847	744	16471	1245	3309	250
1¾	4.4	4049	306	12397	937	20744	1568	4174	316
2	5.0	4924	372	15096	1141	25293	1912	5099	385
2¼	5.6	5848	442	18045	1364	30242	2286	6098	461
2½	6.3	6808	515	21094	1595	35291	2668	7148	540
2¾	6.9	7848	593	24244	1833	40739	3080	8248	624
3	7.5	8888	672	27593	2086	46288	3499	9347	707
3¼	8.1	9987	755	31192	2358	52336	3957	10547	797
3½	8.8	11097	839	34741	2626	58334	4410	11797	892
3¾	9.4	12247	926	38440	2906	64583	4882	13047	986
4	10.0	13446	1017	42289	3197	71181	5381	14396	1088
4¼	10.6	14676	1110	46238	3496	77929	5891	15796	1194
4½	11.3	15896	1202	50287	3802	84777	6409	17245	1304

HEAD		WEIR LENGTH							
		30 cms		90 cms		150 cms		Every 30 cms above 150 cms	
Inches	cms	GPH	LPM	GPH	LPM	GPH	LPM	GPH	LPM
4¾	11.9	17195	1300	54535	4123	91725	6934	18695	1413
5	12.5	18495	1398	58734	4440	99223	7501	20245	1530
5¼	13.1	19770	1495	63083	4769	106471	8049	21694	1640
5½	13.8	21074	1593	67582	5109	114069	8624	23244	1757
5¾	14.4	22444	1697	72081	5449	121967	9221	24743	1871
6	15.0	23819	1801	76729	5801	129965	9825	26393	1995
6¼	15.63			81578	6167	137963	10430	27992	2116
6½	16.25			87077	6583	145961	11035	29792	2252
6¾	16.88			91276	6900	154659	11692	31492	2381
7	17.5			96374	7286	162956	12319	33391	2524
7¼	18.13			101423	7668	171754	12985	35066	2651
7½	18.75			106471	8049	180402	13638	36790	2781
7¾	19.38			111870	8457	189199	14303	38690	2925
8	20			116869	8835	197747	14950	40689	3076
8¼	20.63			122067	9228	206945	15645	42489	3212
8½	21.25			126966	9599	215542	16295	44488	3383

HEAD		WEIR LENGTH							
Inches	cms	30 cms		90 cms		150 cms		Every 30 cms above 150 cms	
		GPH	LPM	GPH	LPM	GPH	LPM	GPH	LPM
8¾	21.88			132764	10037	225490	17047	46438	3511
9	22.5			138213	10449	234887	17757	48487	3666
9¼	23.13			143761	10868	244884	18513	50536	3821
9½	23.75			149210	11280	254832	19265	52536	3972
9¾	24.38			155008	11719	264329	19983	54535	4123
10	25			160757	12153	274426	20747	56785	4293
10½	26.25			173953	13151	296920	22447	61484	4648
11	27.5			185750	14043	317665	24015	65982	4988
11½	28.75			197947	14965	338909	25622	70481	5328
12	30			209194	15815	358154	27076	74730	5650
12½	31.25			221441	16741	379648	28701	78729	5952
13	32.5			232938	17610	400393	30270	82976	6727
13½	33.75			247434	18706	425386	32159	88976	6727
14	35			260680	19707	448880	33935	94225	7123
14½	36.25			273677	20690	471874	35674	99223	7501
15	37.5			286923	21691	495867	37488	104472	7898

HEAD		WEIR LENGTH							
Inches	cms	30 cms		90 cms		150 cms		Every 30 cms above 150 cms	
		GPH	LPM	GPH	LPM	GPH	LPM	GPH	LPM
15½	38.75			300669	22731	519861	39301	108221	8182
16	40			314416	23770	544854	41191	114969	8692
16½	41.25			328162	24809	568848	43005	120468	9107
17	42.5			346157	26169	598340	45234	125966	9523
17½	43.75			356904	26982	620334	46897	131965	9977
18	45			370401	28002	644827	48749	137213	10373
18½	46.25			384647	29079	670320	50676	142712	10789
19	47.5			398893	30156	696813	52676	148460	11224
19½	48.75			413889	31290	722806	54644	154459	11677
20	50			428885	32424	749133	56634	159832	12083
20½	51.25			444131	33576	775375	58618	165456	12508
21	52.5			459377	34729	801618	60602	171079	12934
21½	53.75			474623	35881	827861	62586	176703	13359

(computed and edited from base data in 'Water well Handbook by Keith E.Anderson) A.R.Mahendra

In case Tables are not available, or a specific quantum of flow is not available in the Table, flow can be computed from the equation:

For 90° triangular notch weir $Q = 2.4381 \times H^{5/2}$

For 60° triangular notch weir $Q = 1.4076 \times H^{5/2}$

For rectangular weirs $Q = 3.33 (L - 0.2H) \times H^{1.5}$

Where:
Q = Flow in cusecs.
L = Length of weir opening in feet(should be 4 to 8 times H)
H = Head of water on the weir, for rectangular weirs, Head of water above apex of notch in feet for triangular weirs.

Measuring flow by manometer and the delivery pipe fitted with orifice plate(Table 14):

Discharging flow through a pipe fitted with an orifice plate and computing quantity by a manometer is a frequent method to measure output from high yielding wells, operating with submersible or turbine pumps. Method is not valid for measuring pulsating flows from piston pumps. Fig. 10 shows the essential features of the assembly. A 3 to 10 mm thick metallic cover or a plate with a rounded orifice in the exact center is welded or fitted to the end of the discharging pipe. Threaded caps screwed over a discharge pipe will also serve the purpose. The diameter of the orifice is one half to three-fourths the diameter of the pipe; and has to run full while in operation. The discharge pipe must be straight and level for a length of minimum 2 m (6ft) upstream of the orifice plate, and is supported suitably by a 'Y". Exactly 61 cms (24 inches) back of the orifice plate the pipe wall is drilled and tapped to create a bore, into which a short 3 or 6mm (1/8 or ¼ inch) nipple is screwed in until the inner end is flush with the inner wall of the pipe so as not to protrude inside the pipe. One end of a piece of rubber hose is slipped over the end of the nipple, and the other end over a transparent plastic tubing, or a glass tube, of 1 to 2 m (3 to 6 ft) length which is supported by an accurate scale marked in cms or inches. As the well starts discharging water, the backpressure created by the orifice, forces water up the manometer. There is a finite relationship between the volume of discharge, pipe diameter, orifice diameter and the height to which water is forced up in the manometer tube. The water level in the glass or plastic tube is directly proportional to the orifice-discharge pipe ratio that varies with the pressure head on the orifice. Water level in the manometer indicates the pressure build-up in the discharge pipe. If the respective diameters of the pipe and the orifice are known, volume of discharge can be either computed or directly read from respective tables (Table 14).

This set-up, when properly made, provides the pumping rate within two percent of the true value. Device is particularly useful where the rate of flow is constantly fluctuating or turbulent.

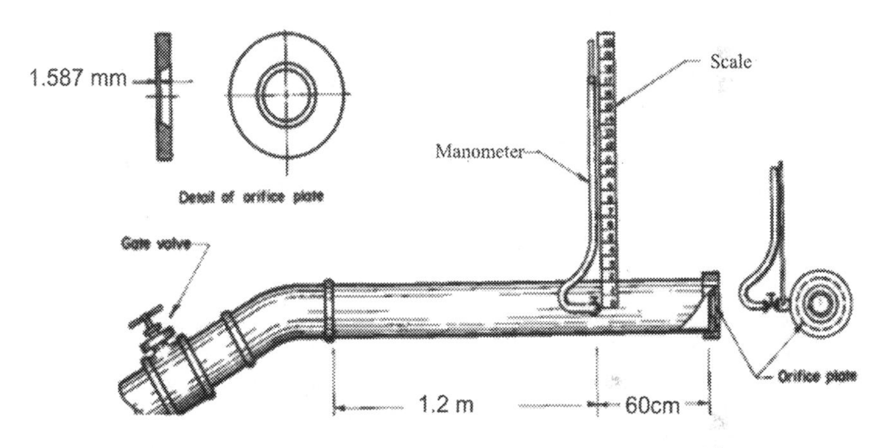

Fig 10: Measuring volume of flow with manometer and orifice plate (Courtesy: Johnson Inc.,: Groundwater and wells)

Table 14 shows the volume of flow in imperial gallons per hour, and in liters per minute, for various combinations of pipes and orifice's of 10 to 25 cms (4 to 10 inches) diameter. With this device, it is possible to measure quantum of flow ranging from 5000 GPH (375 LPM) to 80000 GPH (6000 LPM). The Tables are based on the data compiled by the Engineering Division of Layne and Bowler, incorporated from original calibrations by Purdue University, USA.

Table 14: Measuring flow with manometer and orifice plates

U-tube Reading		3" (7.5cm) orifice				4" (10cm) orifice				5" (12.5cm) orifice			
Inches	cms	4" pipe		6" pipe		6" pipe		8" pipe		6" pipe		8" pipe	
		GPH	LPM	GPH	LPM	GPH	LPM	GPH	LPM	GPH	LPM	GPH	LPM
5	13	5000	378	3800	287	7250	548	7000	529	14000	1058	11000	832
5½	14	5200	393	3950	299	7650	578	7250	548	14650	1108	11500	869
6	15	5400	408	4100	310	8000	605	7500	567	15250	1153	12000	907
6½	16	5550	420	4250	321	8350	631	7750	586	15800	1194	12500	945
7	18	5750	435	4400	333	8600	650	8000	605	16400	1240	13000	983
7½	19	5950	450	4550	344	8950	677	8250	624	16950	1281	13500	1021
8	20	6100	461	4700	355	9250	699	8500	643	17500	1323	14000	1058
8½	21	6250	473	4800	363	9500	718	8750	662	18050	1365	14450	1092
9	23	6400	484	4950	374	9750	737	9000	680	18600	1406	14900	1126
9½	24	6500	491	5100	386	10000	756	9250	699	19150	1448	15350	1160
10	25	6650	503	5200	393	10250	775	9500	718	19650	1486	15800	1194
10½	26	6850	518	5350	404	10500	794	9750	737	20100	1520	16200	1225
11	28	7000	529	5450	412	10750	813	10000	756	20600	1557	16500	1247

U-tube Reading		3" (7.5cm) orifice				4" (10cm) orifice				5" (12.5cm) orifice			
Inches	cms	4" pipe		6" pipe		6" pipe		8" pipe		6" pipe		8" pipe	
		GPH	LPM	GPH	LPM	GPH	LPM	GPH	LPM	GPH	LPM	GPH	LPM
11½	29	7150	541	5550	420	11000	832	10200	771	21050	1591	16900	1278
12	30	7300	552	5700	431	11250	851	10400	786	21500	1625	17300	1308
12½	31	7450	563	5800	438	11500	869	10600	801	21950	1659	17700	1338
13	33	7550	571	5900	446	11700	885	10800	816	22400	1693	18100	1368
13½	34	7700	582	6050	457	11950	903	10950	828	22850	1727	18450	1395
14	35	7850	593	6150	465	12150	919	11200	847	23250	1758	18800	1421
14½	36	7950	601	6300	476	12350	934	11350	858	23650	1788	19150	1448
15	38	8100	612	6400	484	12500	945	11550	873	24000	1814	19500	1474
15½	39	8200	620	6500	491	12700	960	11700	885	24400	1845	19800	1497
16	40	8350	631	6600	499	12850	971	11900	900	24750	1871	20100	1520
16½	41	8500	643	6700	507	13050	987	12050	911	25150	1901	20400	1542
17	43	8600	650	6800	514	13200	998	12250	926	25500	1928	20700	1565
17½	44	8750	662	6900	522	13400	1013	12450	941	25850	1954	21000	1588
18	45	8900	673	7000	529	13550	1024	12600	953	26200	1981	21300	1610
18½	46	9000	680	7100	537	13750	1040	12800	968	26500	2003	21600	1633

U-tube Reading		3" (7.5cm) orifice				4" (10cm) orifice				5" (12.5cm) orifice			
Inches	cms	4" pipe		6" pipe		6" pipe		8" pipe		6" pipe		8" pipe	
		GPH	LPM	GPH	LPM	GPH	LPM	GPH	LPM	GPH	LPM	GPH	LPM
19	48	9150	692	7200	544	13900	1051	12950	979	26800	2026	21900	1656
19½	49	9250	699	7300	552	14100	1066	13150	994	27100	2049	22200	1678
20	50	9350	707	7400	559	14250	1077	13300	1005	27400	2071	22450	1697
20½	51	9500	718	7500	567	14450	1092	13500	1021	27700	2094	22750	1720
21	53	9600	726	7600	575	14600	1104	13650	1032	28000	2117	23000	1739
21½	54	9750	737	7700	582	14750	1115	13750	1040	28300	2139	23250	1758
22	55	9850	745	7800	590	14950	1130	13950	1055	28600	2162	23500	1777
22½	56	9950	752	7900	597	15100	1142	14100	1066	28900	2185	23750	1796
23	58	10050	760	8000	605	15250	1153	14250	1077	29200	2208	23950	1811
23½	59	10150	767	8100	612	15350	1160	14400	1089	29500	2230	24200	1830
24	60	10250	775	8200	620	15500	1172	14550	1100	29800	2253	24400	1845
24½	61	10350	782	8250	624	15700	1187	14700	1111	30100	2276	24600	1860
25	63	10500	794	8350	631	15850	1198	14850	1123	30400	2298	24800	1875
25½	64	10600	801	8450	639	16000	1210	15000	1134	30700	2321	25000	1890
26	65	10700	809	8550	646	16150	1221	15150	1145	30100	2344	25200	1905

U-tube Reading		3" (7.5cm) orifice				4" (10cm) orifice				5" (12.5cm) orifice			
Inches	cms	4" pipe		6" pipe		6" pipe		8" pipe		6" pipe		8" pipe	
		GPH	LPM	GPH	LPM	GPH	LPM	GPH	LPM	GPH	LPM	GPH	LPM
26½	66	10800	816	8650	654	16300	1232	15250	1153	31300	2366	25400	1920
27	68	10950	828	8700	658	16450	1244	15400	1164	31600	2389	25600	1935
27½	69	11050	835	8800	665	16600	1255	15550	1176	31900	2412	25800	1950
28	70	11100	839	8850	669	16750	1266	15700	1187	32200	2434	26000	1966
28½	71	11200	847	8950	677	16850	1274	15850	1198	32500	2457	26200	1981
29	73	11300	854	9000	680	17000	1285	16000	1210	32800	2480	26400	1996
29½	74	11400	862	9100	688	17150	1297	16150	1221	33100	2502	26600	2011
30	75	11500	869	9150	692	17300	1308	16250	1229	33400	2525	26800	2026
30½	76	11600	877	9250	699	17400	1315	16400	1240	33700	2548	27000	2041
31	78	11750	888	9300	703	17550	1327	16500	1247	34000	2570	27200	2056
31½	79	11800	892	9400	711	17700	1338	16650	1259	34300	2593	27400	2071
32	80	11950	903	9450	714	17850	1349	16750	1266	34600	2616	27600	2087
32½	81	12000	907	9550	722	18000	1361	16900	1278	34850	2635	27800	2102
33	83	12100	915	9600	726	18150	1372	17000	1285	35150	2657	28000	2117
33½	84	12200	922	9700	733	18300	1383	17100	1293	35450	2680	28200	2132

U-tube Reading		3" (7.5cm) orifice				4" (10cm) orifice				5" (12.5cm) orifice			
Inches	cms	4" pipe		6" pipe		6" pipe		8" pipe		6" pipe		8" pipe	
		GPH	LPM	GPH	LPM	GPH	LPM	GPH	LPM	GPH	LPM	GPH	LPM
34	85	12300	930	9750	737	18450	1395	17250	1304	35750	2703	28400	2147
34½	86	12400	937	9800	741	18600	1406	17350	1312	36000	2722	28600	2162
35	88	12500	945	9850	745	18750	1418	17450	1319	36300	2744	28800	2177
35½	89	12600	953	9900	748	18850	1425	17550	1327	36600	2767	29000	2192
36	90	12700	960	10000	756	19000	1436	17700	1338	36850	2786	29200	2208
36½	91	12800	968	10050	760	19150	1448	17800	1346	37150	2809	29400	2223
37	93	12850	971	10150	767	19250	1455	17900	1353	37400	2827	29600	2238
37½	94	12950	979	10200	771	19400	1467	18000	1361	37700	2850	29800	2253
38	95	13000	983	10250	775	19500	1474	18150	1372	37950	2869	30000	2268
38½	96	13100	990	10300	779	19650	1486	18250	1380	38250	2892	30200	2283
39	98	13150	994	10400	786	19800	1497	18350	1387	38500	2911	30400	2298
39½	99	13250	1002	10450	790	19900	1504	18450	1395	38800	2933	30600	2313
40	100	13300	1005	10500	794	20050	1516	18550	1402	39050	2952	30800	2328
40½	101	13350	1009	10550	798	20150	1523	18650	1410	39300	2971	31000	2344
41	103	13450	1017	10600	801	20300	1535	18750	1418	39500	2986	31200	2359

U-tube Reading		3" (7.5cm) orifice				4" (10cm) orifice				5" (12.5cm) orifice			
Inches	cms	4" pipe		6" pipe		6" pipe		8" pipe		6" pipe		8" pipe	
		GPH	LPM	GPH	LPM	GPH	LPM	GPH	LPM	GPH	LPM	GPH	LPM
41 ½	104	13550	1024	10650	805	20400	1542	18900	1429	39750	3005	31400	2374
42	105	13600	1028	10700	809	20550	1554	19000	1436	40000	3024	31550	2385
42 ½	106	13700	1036	10800	816	20650	1561	19100	1444	40250	3043	31750	2400
43	108	13750	1040	10850	820	20750	1569	19200	1452	40500	3062	31900	2412
43 ½	109	13850	1047	10900	824	20900	1580	19300	1459	40750	3081	32100	2427
44	110	13900	1051	10950	828	21000	1588	19400	1467	41000	3100	32250	2438
44 ½	111	14000	1058	11000	832	21100	1595	19500	1474	41200	3115	32450	2453
45	113	14050	1062	11100	839	21250	1607	19600	1482	41400	3130	32600	2465
45 ½	114	14150	1070	11150	843	21350	1614	19700	1489	41600	3145	32800	2480
46	115	14200	1074	11200	847	21450	1622	19800	1497	41850	3164	32950	2491
46 ½	116	14250	1077	11250	851	21600	1633	19950	1508	42100	3183	33150	2506
47	118	14350	1085	11350	858	21700	1641	20050	1516	42350	3202	33300	2517
47 ½	119	14450	1092	11400	862	21850	1652	20450	1546	42550	3217	33450	2529
48	120	14500	1096	11450	866	22000	1663	20250	1531	42750	3232	33600	2540
48 ½	121	14600	1104	11500	869	22100	1671	20350	1538	42950	3247	33800	2555

U-tube Reading		3" (7.5cm) orifice				4" (10cm) orifice				5" (12.5cm) orifice			
Inches	cms	4" pipe		6" pipe		6" pipe		8" pipe		6" pipe		8" pipe	
		GPH	LPM	GPH	LPM	GPH	LPM	GPH	LPM	GPH	LPM	GPH	LPM
49	123	14650	1108	11550	873	22200	1678	20450	1546	43150	3262	33950	2567
49 ½	124	14700	1111	11600	877	22300	1686	20550	1554	43400	3281	34150	2582
50	125	14800	1119	11700	885	22400	1693	20650	1561	43600	3296	34300	2593
50½	126	14900	1126	11750	888	22500	1701	20750	1569	43800	3311	34500	2608
51	128	15000	1134	11800	892	22650	1712	20850	1576	44000	3326	34650	2620
51½	129	15050	1138	11850	896	22750	1720	20950	1584	44200	3342	34850	2635
52	130	15100	1142	11900	900	22850	1727	21050	1591	44400	3357	35000	2646
52½	131	15150	1145	11950	903	22950	1735	21150	1599	44600	3372	35200	2661
53	133	15200	1149	12000	907	23050	1743	21250	1607	44800	1187	35350	2672
53 ½	134	15250	1153	12050	911	23150	1750	21350	1614	45000	3402	35550	2688
54	135	15350	1160	12150	919	23250	1758	21450	1622	45200	3417	35700	2699
54¼	136	15450	1168	12200	922	23350	1765	21550	1629	45400	3432	35900	2714
55	138	15500	1172	12300	930	23450	1773	21650	1637	45600	3447	36050	2725
55½	139	15550	1176	12350	934	23550	1780	21750	1644	45750	3459	36250	2741
56	140	15650	1183	12400	937	23600	1784	21850	1652	45950	3474	36350	2748

U-tube Reading		3" (7.5cm) orifice				4" (10cm) orifice				5" (12.5cm) orifice			
		4" pipe		6" pipe		6" pipe		8" pipe		6" pipe		8" pipe	
Inches	cms	GPH	LPM	GPH	LPM	GPH	LPM	GPH	LPM	GPH	LPM	GPH	LPM
56½	141	15700	1187	12450	941	23700	1792	21950	1659	46150	3489	36500	2759
57	143	15750	1191	12500	945	23800	1799	22050	1667	46350	3504	36650	2771
57 ½	144	15800	1194	12550	949	23900	1807	22150	1675	46500	3515	36800	2782
58	145	15850	1198	12600	953	24000	1814	22250	1682	46700	3531	36950	2793
58½	146	15950	1206	12650	956	24100	1822	22350	1690	46900	3546	37100	2805
59	148	16000	1210	12700	960	24250	1833	22450	1697	47100	3561	37250	2816
59½	149	16050	1213	12800	968	24350	1841	22550	1705	47250	3572	37450	2831
60	150	16150	1221	12850	971	24450	1848	22650	1712	47400	3583	37550	2839
60½	151	16200	1225	12900	975	24550	1856	22750	1720	47550	3595	37700	2850
61	153	16250	1229	12950	979	24600	1860	22850	1727	47750	3610	37850	2861
61½	154	16300	1232	13050	987	24700	1867	22950	1735	47900	3621	38000	2873
62	155	16400	1240	13100	990	24800	1875	23050	1743	48050	3633	38150	2884
62½	156	16450	1244	13150	994	24900	1882	23150	1750	48200	3644	38300	2895
63	158	16500	1247	13200	998	25000	1890	23250	1758	48400	3659	38450	2907
63½	159	16550	1251	13250	1002	25100	1898	23350	1765	48550	3670	38600	2918

U-tube Reading		3" (7.5cm) orifice				4" (10cm) orifice				5" (12.5cm) orifice			
Inches	cms	4" pipe		6" pipe		6" pipe		8" pipe		6" pipe		8" pipe	
		GPH	LPM	GPH	LPM	GPH	LPM	GPH	LPM	GPH	LPM	GPH	LPM
64	160	16650	1259	13300	1005	25200	1905	23450	1773	48700	3682	38750	2930
64½	161	16700	1263	13350	1009	25350	1916	23550	1780	48850	3693	38900	2941
65	163	16750	1266	13400	1013	25450	1924	23600	1784	49050	3708	39050	2952
65½	164	16800	1270	13450	1017	25550	1932	23700	1792	49200	3720	39200	2964
66	165	16900	1278	13550	1024	25650	1939	23750	1796	49400	3735	39350	2975
66½	166	16950	1281	13600	1028	25750	1947	23850	1803	49550	3746	39500	2986
67	168	17000	1285	13650	1032	25850	1954	23950	1811	49750	3761	39650	2998
67½	169	17050	1289	13700	1036	25900	1958	24050	1818	49900	3772	39800	3009
68	170	17150	1297	13750	1040	26000	1966	24150	1826	50100	3788	39950	3020
68½	171	17200	1300	13800	1043	26050	1969	24250	1833	50250	3799	40100	3032
69	173	17300	1308	13850	1047	26150	1977	24350	1841	50450	3814	40250	3043
69½	174	17350	1312	13900	1051	26200	1981	24450	1848	50600	3825	40400	3054
70	175	17450	1319	14000	1058	26250	1985	24550	1856	50800	3840	40550	3066

(Computed and edited from base data in 'Water well Handbook by Keith E.Anderson)

Ready, pre-calculated figures available as' Tables' are by far the most convenient way of measuring flow by the orifice method. However, if the Tables are not available, the flow can be computed from the following equation, which involves a protracted and time-consuming procedure:

$$Q = A \times V \times C$$

Where:

Q = Flow per unit of time
A = Area of the orifice
V = Velocity of flow through the orifice
C = Coefficient of discharge at the orifice

The velocity of water as it discharges through the orifice is its velocity in the pipe, plus the velocity gained, as a result of the pressure drop between the point where the pressure is being measured by the manometer tube, and the exit from the orifice. Since the flow is exiting at atmospheric pressure, the head reflected in the manometer converts to velocity, neglecting friction in the pipe.

Estimating volume of flow from self-flowing wells (Table 15):

The approximate flow from vertical pipes (or casings), as in artesian or self-flowing wells can be estimated by collating the diameter of the pipe and the height to which the water is jetting above the pipe. The procedure can also be adapted to measure flow from pumping wells, when the discharge pipe can be turned upward. The vertical pipe should be straight and not less than 1 meter in length, so that the open end is that far above the nearest elbow, bend, or valve. Fig.11 illustrates schematically the general procedure to measure height of jet or flow over a vertical pipe when the height of water flowing over the pipe rim is small the discharge simulates flow over a weir; (F.E.Lawrence and P.L.Braunworth, Cornell University 1906), and when the flow is jetting appreciably over the open end the discharge is computed by an equation governing such flow:

$$\text{Imp GPH} = 284 \times C \times d^2 \times H$$
$$\text{LPH} = 62.5 \times C \times d^2 \times H$$

Where:

d = Inside diameter of the pipe in inches
H = Height of water jet in inches
C = A constant varying from 0.87 to 0.97 for pipes of 5 to 15 cms in diameter and heights from 15 to 60 cms above the top of casing.

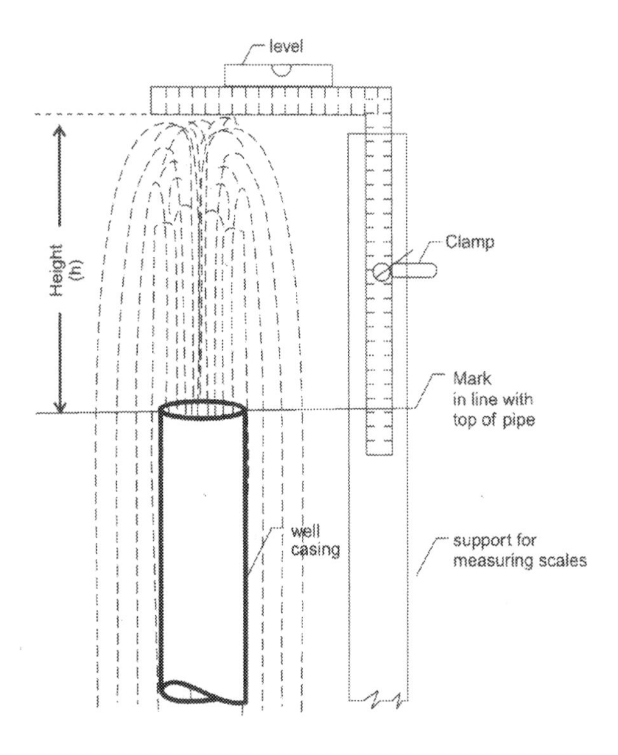

Fig 11: Estimating volume of flow in self-flowing wells and vertical casings
(figure modified courtesy of 'Groundwater and wells'—Johnson inc.)

Table 15 provides the base data to estimate flow from artesian and self-flowing wells, or water-jet from pumped wells with vertical discharge pipes or casings, for various stream heights and pipe diameters both in imperial gallons per hour (GPH) and liters per minute (LPM).

Table 15: Volume of water overflowing from self-flowing wells and vertical pipes

Head of overflow		Nominal diameter of pipe							
		2"(5cms)		3"(7.5cms)		4"(10cms)		5"(12.5cms)	
Inches	cms	GPH	LPM	GPH	LPM	GPH	LPM	GPH	LPM
1½	4	1100	83	2150	163	3400	257	4250	321
2	5	1300	98	2750	208	4650	352	6000	454
3	8	1700	129	3700	280	6600	499	9750	737
3½	9	1900	144	4250	321	7500	567	11750	888
4	10	2050	155	4500	340	8000	605	13250	1002
4 ½	11	2200	166	4900	370	8600	650	13750	1040
5	13	2300	174	5100	386	9000	680	14200	1074
5½	14	2450	185	5450	412	9500	718	15000	1134
6	15	2550	193	5600	423	10000	756	15800	1194
6½	17	2700	204	6000	454	10500	794	16500	1240
7	18	2850	215	6300	476	11000	832	17250	1304
8	20	3000	227	6500	491	11500	869	18500	1399
9	23	3100	234	7000	529	12500	945	19800	1497
10	25	3250	246	7500	567	13000	983	20750	1569
12	30	3550	268	8000	605	14500	1096	23000	1739
14	36	3800	287	9000	680	16000	1210	25000	1890
16	41	41450	314	9750	737	17100	1293	27000	2041
18	46	4500	340	10250	775	18200	1376	28500	2155
20	51	5000	378	11000	832	19300	1459	30000	2268
25	64	5600	423	12000	907	21650	1637	34000	2570
30	76	6200	469	13500	1021	23800	1799	37000	2797
35	89	6750	510	15000	1134	25800	1950	40500	3062
40	102	7250	548	16000	1210	27550	2083	43250	3270

(computed and edited from 'Water well Handbook by Keith E.Anderson)

Estimating flow in horizontal and inclined pipes (Table 16):

Volume of discharge from a horizontal or inclined pipe flowing full, with free fall from its end can be estimated, by measuring the horizontal and vertical distances from the end of the pipe to a point in the stream trajectory as shown in fig. 12. Measure the distance from the point of exit of flow from the pipe end, to a point where the total vertical plunge is 30 cms (12"), either with a carpenter's right-angled scale or two ordinary scales held perpendicular to each other. In case the falling stream is tending to splay, the point of vertical measurement may be located in the middle of the stream, but the vertical measurement should be made from the top of the stream where it exits from the pipe. Small flows of less than 38 LPM (500 GPH) cannot be estimated by this method. Table16 gives the volume of flow, flowing full in horizontal or inclined pipes of diverse diameters, (in GPH and LPM) measured by this means.

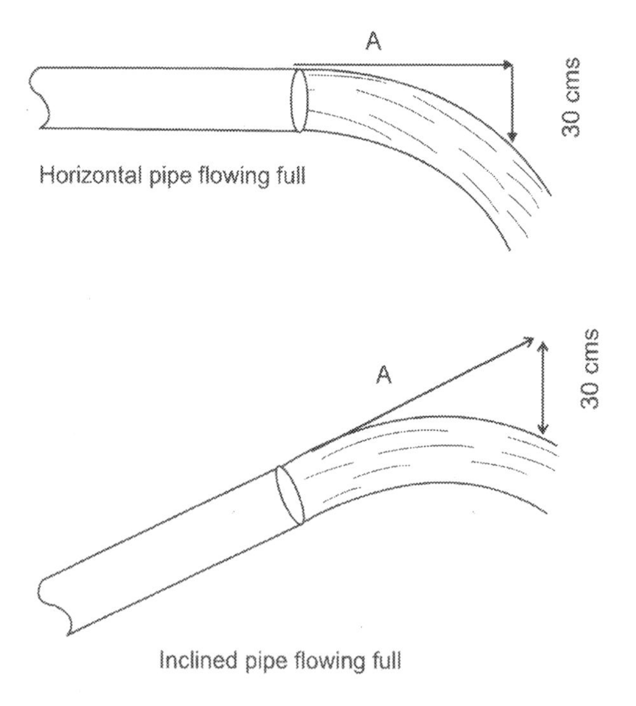

Fig 12: Estimating volume of flow from horizontal and inclined pipes

Table 16: Volume of flow in horizontal and inclined pipes *(Courtesy: U.S. Geological Survey)*

Length 'A' at 30cm plunge		Pipe inner diameter									
		2"(5cms)		2.5"(6.25cms)		3"(7.5cms)		3.5"(8.75cms)		4"(10cms)	
Inches	cms	GPH	LPM	GPH	LPM	GPH	LPM	GPH	LPM	GPH	LPM
4	10	650	49	1020	77	1460	110	2000	151	2620	198
5	12.5	850	64	1275	96	1825	138	2500	189	3275	248
6	15	1000	76	1530	116	2190	166	3000	227	3930	297
7	17.5	1150	87	1785	135	2555	193	3500	265	4585	347
8	20	1300	98	2040	154	2920	221	4000	302	5240	396
9	22.5	1500	113	2295	174	3285	248	4500	340	5895	446
10	25	1650	125	2550	193	3650	276	5000	378	6550	495
11	27.5	1800	136	2805	212	4015	304	5500	416	7205	545
12	30	2000	151	3060	231	4380	331	6000	454	7860	594
13	32.5	2150	163	3315	251	4745	359	6500	491	8515	644
14	35	2300	174	3570	270	5110	386	7000	529	9170	693
15	37.5	2500	189	3825	289	5475	414	7500	567	9825	743
16	40	2650	200	4080	308	5840	442	8000	605	10480	792
17	42.5	2800	212	4335	328	6205	469	8500	643	11135	842
18	45	2950	223	4590	347	6570	497	9000	680	11790	891

Length 'A' at 30cm plunge		Pipe inner diameter									
		2"(5cms)		2.5"(6.25cms)		3"(7.5cms)		3.5"(8.75cms)		4"(10cms)	
Inches	cms	GPH	LPM	GPH	LPM	GPH	LPM	GPH	LPM	GPH	LPM
19	47.5	3150	238	4845	366	6935	524	9500	718	12445	941
20	50	3300	249	5100	386	7300	552	10000	756	13100	990
21	52.5	3450	261	5355	405	7665	579	10500	794	13755	1040
22	55	3650	276	5610	424	8030	607	11000	832	14410	1089
23	57.5	3800	287	5865	443	8395	635	11500	869	15065	1139
24	60	3950	299	6120	463	8760	662	12000	907	15720	1188

Length 'A' at 30cms plunge		Pipe inner diameter							
		5"(12.5cms)		6"(15cms)		7"(17.5cms)		8"(20cms)	
Inches	cms	GPH	LPM	GPH	LPM	GPH	LPM	GPH	LPM
4	10	4080	308	5880	445	8000	605	10460	791
5	12.5	5100	386	7350	556	10000	756	13075	988
6	15	6120	463	8820	667	12000	907	15690	1186
7	17.5	7140	540	10290	778	14000	1058	18305	1384
8	20	8160	617	11760	889	16000	1210	20920	1582

Length ‘A’ at 30cms plunge		Pipe inner diameter							
		5"(12.5cms)		6"(15cms)		7"(17.5cms)		8"(20cms)	
Inches	cms	GPH	LPM	GPH	LPM	GPH	LPM	GPH	LPM
9	22.5	9180	694	13230	1000	18000	1361	23535	1779
10	25	10200	771	14700	1111	20000	1512	26150	1977
11	27.5	11220	848	16170	1222	22000	1663	28765	2175
12	30	12240	925	17640	1334	24000	1814	31380	2372
13	32.5	13260	1002	19110	1445	26000	1966	33995	2570
14	35	14280	1080	20580	1556	28000	2117	36610	2768
15	37.5	15300	1157	22050	1667	30000	2268	39225	2965
16	40	16320	1234	23520	1778	32000	2419	41840	3163
17	42.5	17340	1311	24990	1889	34000	2570	44455	3361
18	45	18360	1388	26460	2000	36000	2722	47070	3558
19	47.5	19380	1465	27930	2112	38000	2873	49685	3756
20	50	20400	1542	29400	2223	40000	3024	52300	3954
21	52.5	21420	1619	30870	2334	42000	3175	54915	4152
22	55	22440	1696	32340	2445	44000	3326	57530	4349
23	57.5	23460	1774	33810	2556	46000	3478	60145	4547
24	60	24480	1851	35280	2667	48000	3629	62760	4745

(computed and edited from base data in ‘Water well Handbook’ by Keith E.Anderson)

Estimating volume of discharge in partially flowing pipes (Table17):

To estimate flow in pipes, which are not running full, a reduction factor obtained from Table17 is introduced into the value of flow in pipes running full. Reduction factor is the ratio of freeboard 'F' and the inside diameter 'D' in percentage. Freeboard (F), and inside diameter (D), are measured as shown in fig.13 and the percentage of the ratio F/D is read against the reduction factor. Determine volume of flow for full pipes from Table 16, and this quantum is multiplied by the reduction factor shown against F/D percent in Table17 to obtain the volume of flow in partially flowing pipes. The actual flow in partially flowing pipes will be approximately the value for full flowing pipe of the same diameter multiplied by the correction factor in Table 17.

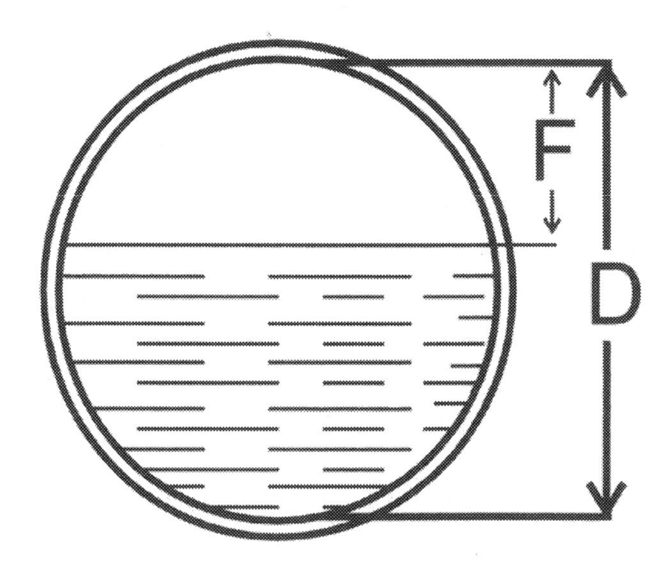

Fig 13: Ratio of diameter of the pipe and the freeboard in partially flowing pipes

Table 17: Correction factors for partially flowing pipes

F/D %	Factors
5	0.951
10	0.948
15	0.905
20	0.858
25	0.805
30	0.747
35	0.688
40	0.627
45	0.564
50	0.500
55	0.436
60	0.375
65	0.312
70	0.253
75	0.195
80	0.142
85	0.095
90	0.062
95	0.019
100	0.000

(Courtesy: U.S.Geological Survey)
(computed and edited from base data in 'Water well Handbook' by Keith E.Anderson)

CHAPTER IX

TESTING WELLS FOR YIELD & DRAWDOWN

One of the important and significant steps in groundwater extraction is testing a newly drilled well for its dependable yield, and the well's potential to sustain the output over several years. Two types of tests are generally carried out. One is the well test, and the other the aquifer performance test. For general purpose wells, which includes 80% of all the wells drilled, a simple well test is adequate to commission the well rationally, and pump water from it. Selecting a pump at random, and installing it without a yield test will often result in either under-pumping or over-pumping the well. Under pumping a well implies drawing less water than what the well is capable of yielding, leading to wastage of the investment on the well, and spiraling power costs. Example of under pumping is, installing a High-Head pump, in a well where water is at moderate depths. Such a pump will yield less water, even if the well has the prospect to yield higher quantities of water. A High Head pump will lift water from great depths, but will pump proportionately less volume of water. Over-pumping a well means extracting water in excess of the quantity the well is able to yield, which will result in sporadic yield, frequent cessation of the output from the well, 'drying-up' or loss of yield during summer months and eventual failure of the well.

Installing a pump at random, such as a 'low Head—high yield pump' may pump large quantities of water, times more than the capacity of the well. Unless the well has a high output capacity, most of the wells, which are thus over-pumped, will dry up within a course of a year or two. Over-pumping a well, without a well test also has perilous consequences of depleting water levels in a well field.

Aquifer performance tests:

These tests have certain advantages and disadvantages. They are conducted during well field development to acquire aquifer data, over a large area. Aquifer performance tests entail an elaborate set-up, skilled work force, construction of non-pumping observation wells and complex computations. Observation wells, drilled at various distances from the main well react to pumping in the main well, the depletion in their water levels directly reflecting the spread of the cone of depression. These tests are normally appropriate to sedimentary aquifers, which transmit vast quantities of water and the tests form an integral pre-requisite for development of well fields for supplying water to a town water supply system, irrigation, or for an industrial undertaking. From an APT, it is possible to obtain the 'transmissibility-quotient', 'Specific yield' and 'storage coefficient' of the formations from which the well is drawing water. By projecting this data, it is possible to estimate the properties of the aquifers for the well field, and plan an array of wells for the project. These tests are conducted and data analyzed under a host of presumptions, important of which are that the aquifers being tested are, to the extent of their occurrence, are homogeneous and isotropic, are not bound by any boundary conditions, and are not being recharged during the duration of the test. This is rarely the case in nature. Post-testing, except for the main pumping well, the other observation wells, in course of time become irrelevant.

Alternative method is to compute 'T' values by integration of data from mechanical analysis of carefully collected drill cuttings, log data, and yield records, supplemented by laboratory analysis of drill core samples. Various workers have expressed that if accurate well logs are available the transmissivity can be estimated with reasonable accuracy, and that while investigating large groundwater basins, this geological approach is a better option than conducting a large number of expensive pumping tests. (Kruseman, G.P. and N.A. de Ridder, 1994).

This geologic approach involves collection of samples from drill cuttings, preparation of accurate well logs and lithological description of samples from the different geological formations penetrated during drilling. Classified samples are mechanically analyzed in the laboratory to define the grain sizes of the various formations. Field description and numbering of the samples, integrated with the sieve analysis data will define the boundaries between the various formations. In the process, the average grain size, the degree of sorting and the clay and silt content of the sand samples are also categorized for each of the aquifers encountered by the drilled hole. All these parameters will bear on the hydraulic conductivity of the specific formation in various ways. Core and bulked samples tested in a permeameter in the laboratory will also provide permeability data, which together with the well logs is made use of to define the aquifer characteristics. By these studies, a correlation is established between the permeability and the grain size. While establishing this relationship, effects of sorting and the clay and gravel content in the sample is also taken into account (De Ridder and Wit, 1965). Graphs or tables showing the relationship between grain size and permeability are drawn to obtain a cohesive picture of the variations in the aquifer. Permeability coefficient averaged from each of the layers, multiplied by the thickness of the aquifer gives its transmissivity at the well site.

The option of either preferring an aquifer performance test, or a laboratory oriented study of the drill samples to determine transmissivity should be decided based on the purpose for the investigation.

Compared to an APT, a well test is simple. It is conducted on a single well, which eventually converts to a production well providing the required water. Well test described in this chapter, are relevant to single wells drilled in both soft and hard rocks. Even in a single well test, drill samples are systematically collected for laboratory analysis. Following completion of drilling, and before installing a permanent pump, the well is tested, with a test pump, to determine the optimum yield capacity, and interconnected water levels in the well. Many times the test pump itself, if carefully selected, taking into account all particulars during drilling, may suit the well as a permanent production unit. However, most of the times it would be necessary to replace the "test pump" with a permanent pump. The well test provides valuable data revealing the quantum of water that could be safely pumped, and most important, the characteristics of the permanent pump that needs to be installed. The cost of a well test, in time and economy, made prior to purchase of a pump, will pay for itself by the savings that accrue from the additional water obtained, and lower power consumption.

The tests also reveal vital facts concerning the groundwater reservoir-data that cannot be obtained by any other method.

In hard rock formations, groundwater flow does not follow any particular regime; it occurs in a dissipated and abstract manner, and is restricted to fracture porosity in the rock. Groundwater transmits in hard rocks through fractures and joints, and since their connectivity is irregular, observation wells, if drilled, rarely react to pumping in the main well. Hence, in hard rock areas, tests and measurement of drawdown and the water level recovery is limited to the main well.

Data required for both the tests are : i) the output during pumping, in LPM or GPH; ii) water levels measured both during pumping and recovery; and iii) the depth down to which the water level will decline during controlled pumping. In aquifer performance tests, flow is not controlled; the well is allowed to drawdown over an extended pumping period, and the rate of drawdown and yield are continuously monitored. For the majority of the groundwater investigations, a well test is adequate to acquire the data required for selection of proper pump and commissioning of a tube well.

This chapter deals only with the well test. Literature dealing with aquifer performance tests is listed in the 'References' for further reading.

Testing a well involves prior planning during the drilling stage itself. Essential information required is the data on the output of water, monitored during drilling, either in LPM or GPH. Potential of a well to yield water with a pump is generally more than the yield achieved by compressed air during drilling. However, particularly in very hard rocks, such as very fine grained granites, there are instances where high water yield experienced during production pumping, diminished in course of time, relative to the output recorded during drilling, and has even ceased altogether within an indefinite span of time. Time span may extend to hours or days of pumping and abstraction. The borehole intercepting a major, but isolated fracture in the rock or encountering a perched water table causes this abrupt or, sometimes gradual cessation in yield. A perched water table is an isolated, large pocket of water that occurs in the vadose zone, resting above, and cut-off from the main water table or aquifer. They form when an impermeable layer of rock or sediment (aquiclude) or a poorly permeable layer (aquitard) occurs above, and cut off from the main water table or aquifer, but below the land surface. Perched water tables occur both in soft and hard rocks. In hard rocks, large extended joints, fractures, and crevices occur isolated from the main fracture porosity. Rain water percolates and collects in such openings, but lack channels for replenishment from adjacent fractures. Presence of perched and disconnected water tables are potentially dangerous, because during exploration, they lead to misleading data about the presence of large quantities of water under pressure, while actually they hold limited quantities of water under confined pressure, and without any access for replenishment. It is difficult to quantitatively estimate the volume of groundwater that is being held in such isolated pockets, unless sub-surface geological data is known or a well test has been conducted. Drilling production wells based on initial yield, which may be sourcing a perched water table may yield large quantities of water for some days of acceptable pumping, with no sign of the approaching failure of the well; but as the water under restricted storage runs out, the output from such a well will rapidly decline, and will soon cease altogether.

A well test conducted on this well would put on alert its impending failure:

1. By the reduction in water yield during the course of the test;
2. By continuous and uncontrollable fluctuations in the discharge pressure,
3. By the rapid rate of drawdown, and
4. An equally misleading rapid recovery in water levels.

Such a well with unrestrained and rapid fall in output should be considered a dud well, and abandoned, instead of expending cost on pump and pipelines. It is mandatory that all such wells that have been abandoned for various reasons are tightly closed and sealed.

Objectives of a well test are:

1. To quantify the amount of water the well is capable of yielding, leading to selection of the proper pump and its placement level.
2. In case the well is a poor or medium yielder, to control its production output to within its potential, so that recharge into the well, and discharge are adequately balanced.
3. To determine the pumping level (the maximum depth down to which the water level will draw-down during extended

pumping) so as to position the submersible pump (inlets in case of a jet pump) below the NPSH value of the production pump. This would ensure that the water level stays within the suction head of the pump during summer months when the water levels decline.

Preliminary information:

Preliminary data required for the test would normally have been recorded during the drilling of the well. Data consists of two sets of information, relevant while setting up the equipment for the test.

1. Flow measurements made with the v-notch, tabulated with depth.
2. Depths at which initial moisture was noticed and depths at which there occurred any increase or decrease in the quantum of discharge.

Test equipment:

1. Test pump: The test pump should have within its 'Head' range, capacity to lift water from a depth of about 15 m to 5 m more than the depth at which the maximum water flow was recorded during drilling. Output capacity of the test pump should be at least 50% more than the maximum discharge recorded during drilling. In the absence of any preliminary data, taking into account the depth at which water was first encountered during drilling, a medium Head (30-50 m), medium yield (100-130 LPM) pump of about 1.5 HP could be deployed for the test.
2. Pipe lines for the permanent installation.
3. 12 mm (1/2 inch) flexible tubing or GI air-line fastened to the main pump column, as a permanent fixture to monitor water levels.
4. Discharge system with pressure gauge and throttle valve.
5. Devices to measure discharge and depth to water (see chapter V).
6. A stopwatch.
7. A tabulated format sheet on a clip board, with 6 columns marked respectively as i) Time, ii) Tape reading at surface, iii)

Tape wet-cut reading, iv) Depth to water; v) Pressure gauge reading, vi) Discharge in liters per minute.

8. A power Generator, in case the well is located away from a main line power source.

Test set-up: (fig.14)

The test pump is installed at a depth approximately 30 m lower than the depth at which the maximum flow was recorded during drilling, viz., if the max flow was recorded at a depth of 50m, the test pump should be placed at 50 + 30 = 80 m depth. The exact depth of the test pump installation should be carefully noted on the clipboard. Along with the pump column, and tied parallel to it, is lowered a 12mm diameter (1/2 inch) G.I. airline, or flexible tubing. The airline would constitute an integral part of the permanent installation to facilitate measuring water levels in future. Determining the water levels is a prognostic tool in identifying the probable reasons for malfunctioning of a well. The airline extends only as deep as the top of the pump assembly, and is open at its termination. The exact length of the airline inserted into the well should be accurately measured and retained as a record. All connections in the airline should be airtight. Top of the pump column is connected to a horizontal discharge pipe on which the manometer and the throttle valve are mounted in sequence, as shown in the figure.

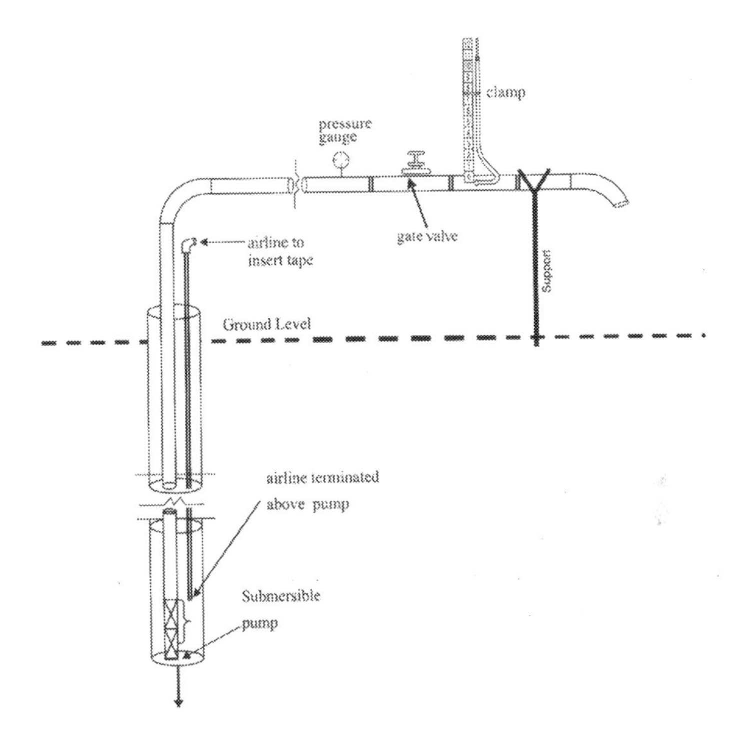

Fig 14: Set-up for a well test; includes systems to measure water levels, discharge pressure and flow rate

Preparations for the test:

1. Static water level (non-pumping level) in the well is accurately measured and recorded.
2. The throttle valve is kept ½ to ¾ ths open, with pressure gauge reading zero.
3. For measuring output, the discharge pipe with the required 'bend' is centered on the measuring drum; any drum of known capacity, viz., 220 lts drum will serve the purpose for low yielding wells. (See chapter on 'measuring water flow'). Discharge measurements could also be made by a flow-meter. A manometer with a synchronized orifice plate is an integral set-up to monitor output from high yielding wells.
4. An electric depth sounder, or a steel tape with chalk rubbed on its surface to a length of about 10m is inserted into the airline, or directly into the well, to a depth of 6m below the static water

level. The tape reading at the top casing head, minus the 'wet cut' (the watermark on the chalked surface) indicates the depth to water level.

Well test:

The primary objective of the well test is to determine the yield capacity of the well, and the drawdown for that discharge, to facilitate selection of an appropriate pump. Procedure involves pumping water, with the throttle valve fully open, under a reasonably constant pressure, and to determine the dependable yield and the maximum drawdown in the well, (which is the 'head' against which the pump will be pumping water). A throttle valve is installed, to control output, in case of poorly yielding wells. In low yielding wells, the throttle valve is maneuvered to stabilize the output and water levels to a constant rate.

1. The pump is switched on; and the exact time of starting the pump is noted against the respective column.
2. Reading on the pressure gauge is monitored. If, and as the gauge tends to record a continuous rise, the valve is opened appropriately to stabilize the reading. Reciprocally, if the gauge tends to recede, the throttle is gently closed to restrain further decline. This opening and closing of the throttle valve is a continuous process until a steady pressure and output are achieved. At regular intervals, pressure gauge readings are noted in the appropriate column.
3. A steady state of flow, with a balanced drawdown, if and when achieved reflects balanced recharge and discharge conditions in the well. A pumping well so stabilized to yield a steady output should be pumped for 12 hours, or as long as possible to confirm its yield capacity. In case the output / pressure tend to decrease, or increase, time of such change is of importance to project the eventual potential of the well. A fully open throttle valve and a steady gauge reading signify a good production well. An erratically behaving pressure gauge indicates rapid decline in water levels reflecting an unbalanced inflow and output. It may become necessary to continuously throttle the flow to maintain pressure, and a steady flow. Such a sequence may also indicate that the yield capacity of the test pump is in excess of the well capacity,

and not adequately calibrated to the discharge recorded during drilling. Insufficient depth of installation of the pump will be indicated by the water levels declining close to the pump location.

4. Concurrent with switching on the pump, water levels, as measured on the tape are recorded against the respective columns. Water level measurements are taken at following time intervals.

   i) During the first 5 minutes at 1 minute intervals
   ii) During the next 10 minutes at 2-minute intervals
   iii) During the next 30 minutes at 5 minute intervals
   iv) During the next 1 hour at 15 minute intervals
   v) For the next 6 hours at 30 minute intervals.

The above time-intervals are only for guidance. Intervals and duration of the test could be conveniently modified according to circumstances.

As the pumping progresses, three sets of figures become available on the clip board.

i) Time, ii) Water levels, and iii) Output from the well.

This data, when integrated will give a provisional capacity of the well, sufficient to select the appropriate pump. However, in addition to the tabular data, a graphical representation of these figures will present a logical depiction of the well behavior, the ground water flow characteristics, and valuable information, which can be projected for other wells in the area.

Graphical information:

In an aquifer performance test, drawdown and recovery measurements are plotted on semi-logarithmic graph, since duration of the test is for extended periods, and to acquire transmissivitty and storage parameters of the aquifer for wider application. Plotting on a semi-log sheet yield straight line graphs, making the computations from data analysis easier. Drawdown between two log cycles inserted into relevant formulae give the transmissivity value of the aquifer. In a well test, measurements can be plotted either on an arithmetic graph or on a semi-log sheet depending upon the duration of the test, with drawdown measurements, in meters, on the ordinate or the vertical (y) scale, and a parallel line showing output in LPM also on the vertical scale. Elapsed time in minutes is plotted on the abscissa or the horizontal (x) axis. Data will be valid for wells, where the yield is good, without any throttling by the gate valve. The asymptotic curve that emerges from the graph will reflect the rate of, and the absolute drawdown levels in real time. The fall will be steep during the initial minutes of pumping; but the curve will gradually and progressively flatten out, assuming a nearly prone attitude, in course of time, with reference to the horizontal axis. This signifies approach of stable, ideal hydraulic conditions, where the inflow into the well just about balances the output. Since the objective of the test is to establish the point in time when the discharge just balances the inflow into the well, the output should be so controlled, that the drawdown curve after the initial steep fall, and flattening out, tends to rise parallel to the horizontal axis. This outline for the plot should be induced, if necessary by repeating the test as explained in step 3, by appropriate closing of the throttle valve to balance recharge and discharge. The curve represents a single point in time, of a constantly enlarging cone of depression. The outer boundaries of the cone of depression represent points from which the pumping well is drawing water.

Abrupt steepening of the curve indicates the cone of depression approaching a dry bounding zone; and a flattening of the curve indicates the depression cone being near a saturated zone.

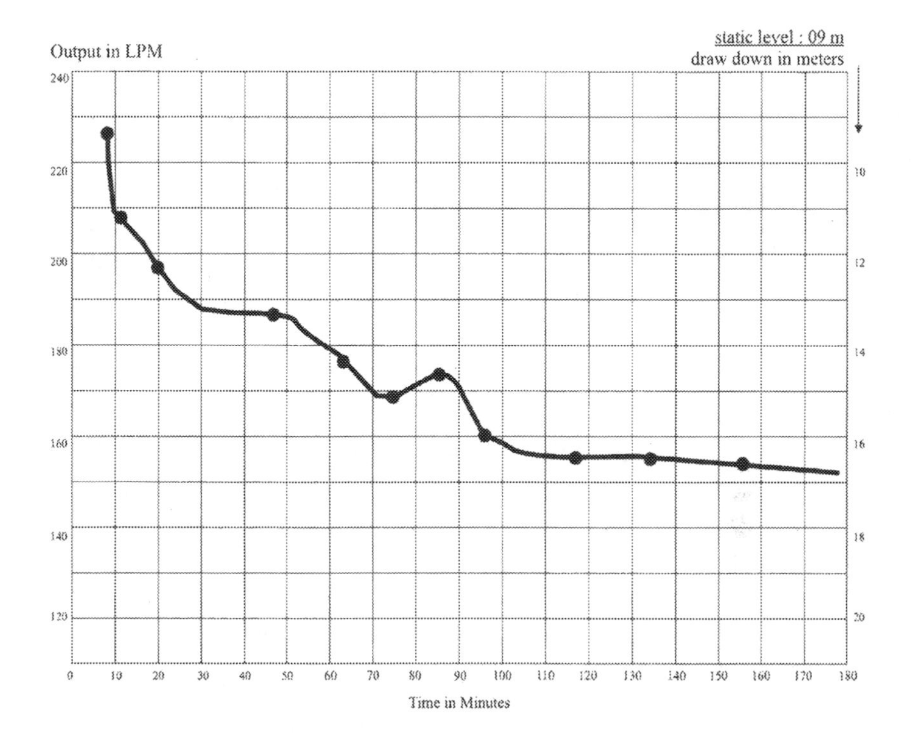

Fig 15: Representative excerpt from a drawdown curve plotted on arithmetic graph

Fig 15 is a condensed graphical representation of a typical well test, plotted on arithmetical graph, conducted on a bore well drilled in highly fractured limestones of the Precambrian Wainganga series, in Orient cement plant, near Mancherial in Adilabad District, Andhra Pradesh. The cement plant, with a production capacity of 3.4 million tonnes per annum, constructed in a totally dry area is entirely dependent on groundwater, both for cement production and for the drinking water needs of a large colony of staff manning the Plant. For similar tests, depending on the static level and the proposed duration of the test, figures on the ordinate and the abscissa will vary, viz., if the drawdown recorded is 50 or 60 m, the 'y' axis should be suitably scaled, and modified to that of the maximum recorded drawdown level. In the test depicted in fig 15, the drawdown data is hazy for the initial 20 minutes after starting pumping. Between 20 and 60 minutes, the drawdown is steep, until the depression cone reaches a water source, and starts drawing water from its confines. At this stage, the falling water levels stabilize and flatten out, till the 74th minute, when they rise indicating the incidence of a major water source. After 85 minutes of pumping, the water levels decline, reflecting the depression cones proximity to an impermeable zone. Two hours after starting pumping, the water levels stabilize at about 17 m below the static level, with a yield of about 155 LPM for a drawdown of 8 m, indicating a higher capacity for the well. Higher capacity pumps will concomitantly increase both yield and drawdown. An extension of the test or a step draw down test will indicate the economical pump and the dependable yield of the well. About 40 wells in the plant area and the surroundings have been tested, and appropriate pumps installed to meet the water requirements of the manufacturing unit and the staff manning the plant.

Conclusions:

It should be possible to conclude from the test the following general performance characteristics of the well:

1. Since the output characteristics of a well with a pump are distinctly different from that obtained during drilling, it should be possible to decide on the dependability of the well for its end application.
2. Results of the well test could also be read from the graph showing the declining water levels and reducing yield with time. The trough and the flat part of the curve, projected to the vertical axis represent the pumping water level, attained with the test pump. This level is the 'Head' against which the pump will be drawing water. The same 'flat' line projected further to the line drawn parallel to the vertical axis will indicate the 'yield' of the well with the test pump, at that point of time. However, any data concluded from the graph will be valid only for the virtual time elapse plotted on the 'x' axis.
3. The permanent production pump should have a Head range approximately +20m and -20m of the recorded pumping level, viz., if the pumping level is 60m below ground level, the production pump should be able to lift water from levels of 60-20 = 40, and 60 + 20 = 80 m levels or equivalent. The pumping level is not absolute, but will vary, taking into account

the seasonal fluctuations in water levels. The required head for the pump should take into account the static head and the dynamic head accruing from i) the distance, ii) height of the discharge point, and iii) the friction head ensuing from the piping system. The graph will have an universal application for this specific well to estimate drawdown and output with different pumps and varying periods of pumping.

4. Yield capacity of the production pump is computed from the discharge recorded at the end of the test, when the curve is tending to rise from its prone attitude of running nearly parallel to the horizontal axis. A flat curve indicates stable pumping conditions, where the input into the well and the output are nearly balanced. For example, if the stabilized, sustained discharge, after extended pumping is 110 LPM (1500 GPH), the yield capacity of the production pump should have 110 LPM as its mid-point, with increased discharge at lower Heads, and decreased discharge at higher Heads (See chapter on pump selection).

Step drawdown test:

A step-drawdown test is a pumping test designed to investigate the performance of the well under controlled, variable discharge conditions. In a step-drawdown test, the discharge rate in the pumping well is increased from an initially low and constant rate through a sequence of pumping intervals (steps) of progressively higher constant rates. Each step is typically of equal duration, lasting from approximately 30 minutes to 2 hours (Kruseman and de Ridder 1994). Step-drawdown tests provide convenient tools for the estimation of the time duration yield of wells, and concurrent drawdown. Test is conducted in 3 or 4 steps, pumping rate being increased at each step, while monitoring the drawdown levels. Time and the rate of drawdown at each of the incremental steps will indicate whether the well is being over-pumped or under-pumped for that particular output. An abrupt increase in the drawdown level, at any pumping step is a pointer that the well cannot sustain that yield. Step at which the drawdown level was stable for a constant discharge signifies the well capacity. Pump selection for the tested well, and the pumping rate should be limited to the output recorded during this step. Provision of a pressure gauge on the discharge arm will help correlate yield, drawdown and the pressure at any particular step. In case an ATP is to be conducted, the data from the step drawdown test will specify the discharge at which the well should be pump tested. In advanced hydrology, step drawdown test facilitates computation of well efficiency, well and aquifer losses.

Measuring water levels:

1. Electrical sounding: An electrical sounder consists of a cable with a pair of insulated positive and negative leads, with open terminals, feeding from a battery. Depths are stamped on the cable which is inserted into the air-line or the pipe. The cable touching the water level is indicated on the surface by an audio-visual signal. Depth inserted marks the water level.
2. Steel tape: A steel measuring tape of sufficient length, weighed down at its '0' terminal, smeared with school chalk to a length of about 3 m, is inserted into the air-line or the pipe, to an estimated depth of the water level in the well. The chalk on the submerged part of the tape shows up as a wetted length, and this 'wet-cut' reading, subtracted from the reading at the surface provides the DTW (Depth to water).
3. Air-line method: Air pressure required to replace water in a submerged tube is equivalent to the pressure exerted by a column of water of corresponding height. This pressure expressed in meters of water (1 kg/cm^2 = 1 m head or 1 $lb/inch^2$ = 2.31 ft) can be converted to depth to water. Head of water in pipe = 1 m × 1 kg/cm^2 or 2.31 × $lbs/inch^2$ as read on the pressure gauge. 1 kg/cm^2 is the pressure exerted by 1 m head of water or 2.31 $lb/inch^2$ pressure by 1 ft head of water. A tire valve in the air-line, attached via a tee with a pressure gauge, to an inflating pump will enable forcing the water out of the line. Reading on the gauge is the pressure required to support the column of water, against the water standing in the well. Gauge reading is converted to meters head of water by the equation 1 kg/cm^2 = 1 m column. This reading shows the submerged length of the airline, which minus the total length of the airline gives the DTW = Depth to water.

CHAPTER X

PUMPING SYSTEMS

The ultimate objective of all groundwater projects is to get water out of the ground where it occurs, and move it to the place of usage. Hence, it is fundamental to understand the various pumping systems, and their economic application. The early beginning of the art of lifting water out of the ground is lost in antiquity, but ability to raise water from its depths is so fundamental, that civilization progressed in tandem with the development of pumping technology. Earliest devices were the shallow to deep domestic wells, with persons drawing water with suitable containers with the help of a winch and straddle. It may appear strange that in spite of all the advances in pumping technology, this method of lifting water is in still in vogue. Wide use of this mode of lifting water is for the reason that it is independent of any energy source. Innumerable techniques have evolved, from the chain-and-bucket based arrangements, to the present day systems which can lift millions of liters of water over heads of hundreds of meters, such as those which operate in high-capacity lift irrigation projects.

The primary action of all pumps is based on the same general principle, where air is initially exhausted from a working chamber to create a near vacuum. Atmospheric pressure acting on the water surface will push it into the air-evacuated chamber. Here it will be picked up by the moving impellers of the pump and forced out of the discharge pipe. A simple method of creating a vacuum is filling the pump casing with water from an external source. When the pump is started, the rotating impeller forces the water out, thus creating a vacuum, and the water rises to fill the vacuum, thus initiating a continuous cycle of suction and discharge. In a reciprocating pump the piston working in the cylinder creates the necessary vacuum. In a hand pump, the plunger moving up and down in the chamber creates the necessary vacuum to draw water.

There are numerous types of pumping systems, each with diverse performance characteristics, designed to suit range of hydraulic situations.

The major systems, in vogue today to lift and move water are:

1. Surface centrifugal pumps
2. Turbine pumps
3. Submersible pumps
4. Jet pumps
5. Reciprocating pumps
6. Rotary pumps
7. Air-lift pumps

All pumps work on the same principle that water rises in a conduit to fill any vacuum. If a pump is to lift water from a well, it is a pre-requisite to create a vacuum, by exhausting air from the working chamber of the pump. Once a partial vacuum is created in the pump's chamber, the atmospheric pressure acting on the water surface will push it into the vacuum. As the water enters the casing through the hub and the intake, the impellers of the pump rotate it around in the volute, imparting a centrifugal force to the water. This pressure will force the water though the discharge pipe.

The air in the chamber is evacuated through an air-vent, by filling it with water from an outside source. Once the pump starts and runs for a while, for subsequent starts, pumps self prime, as the foot valves at the end of the pipe restrain water in the discharge pipe, and the pump chamber. Theoretically, at sea level, the centrifugal pump should draw water to a height of 10 m (34 feet), but rarely does because of many mechanical causes. Dependable height is not more than 6 m, but more frequently 5 meters.

Most of the pumps are based on the principal of centrifugal force, or its variations. In this handbook, the emphasis is on the centrifugal based systems, while the operating principles of other types of pumps find a brief mention.

Surface centrifugal pumps:

A centrifugal pump consists essentially of an impeller, with an intake at its center, so arranged that when it rotates, it will throw water by centrifugal force into the case surrounding the impeller (fig.16). The impeller rotates at 1750 to 3500 RPM; water enters axially though the center or the eye of the impeller, and is discharged radially at its circumference into the volute surrounding it. The volute is a spiral shaped Housing around the impeller.

Fig 16: Cross-sectional image of an impeller

In the modern centrifugal pump, most of the mechanical energy conversion, to move the water, is from the outward force that curved impeller blades impart on the fluid. Concurrently, part of the energy also pushes the fluid into a circular motion, and this circular motion contributes to an increase in the water pressure at the outlet.

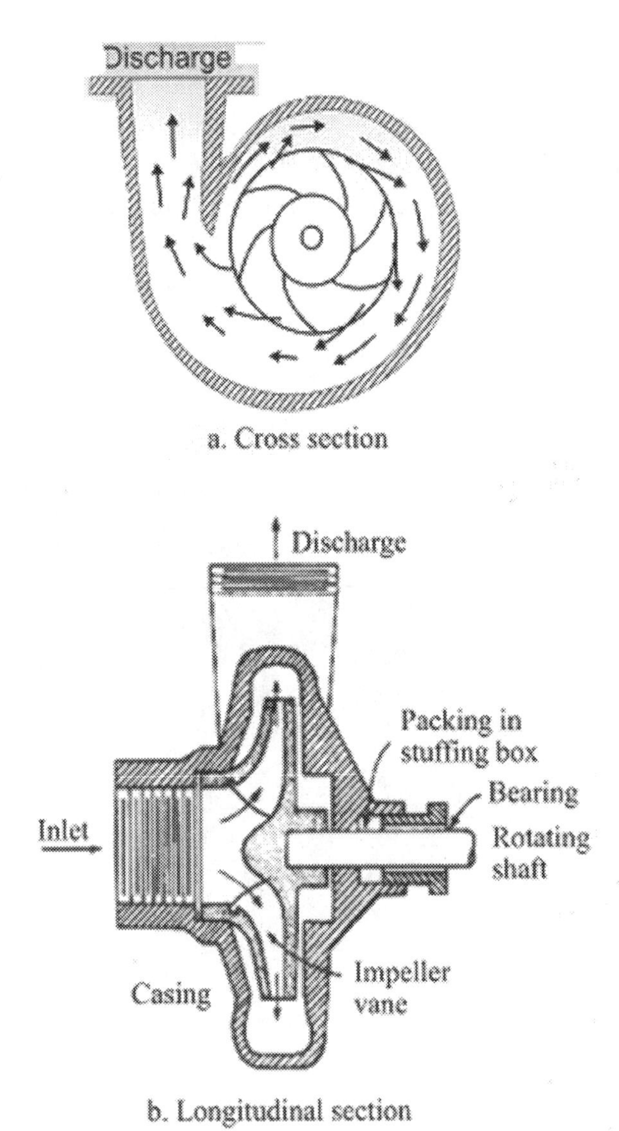

Fig 17: Sectional views of the volute

The volute progressively converts the velocity head created by the fast rotating impeller to a pressure Head to force the flowing water through the discharge pipe. A single stage centrifugal pump has one impeller, while a multi-stage pump has two or more impellers in a single casing, which is designed to lead the discharge from one impeller to the eye or suction of the successive impellers. The travel from one to the successive impeller continuously boosts pressure to the flow.

Most of the pumps in use today, including the submersible and jet pumps are also centrifugal pumps working in different situations and in suitably modified Housings. The main working and maintenance principles for all these pumps are comparable. The centrifugal pump always pumps the difference between the suction and discharge heads; with an integral combination of the head and yield capacity built into the design of the impellers and the volute. These two numbers, Head and output multiplied is a constant. In other words, if the head increases the capacity of the pump will integrally decrease. If the design for the yield increases, the Head against which the machine can pump will decrease reciprocally. This inverse relation of Head and yield is of critical importance in selection of pumps, and consequentially the functioning of the well.

Centrifugal pumps possess following advantages over other types of pumps:

1. A functional speed, which allows direct connection to electric or diesel motors;
2. Compactness and low cost;
3. Smooth flow through the pump, and incidence of uniform pressure in the discharge pipe;
4. Power characteristics that make it an easy load for the prime mover. Any increase in Head reduces the power required, a characteristic that makes overloading of the motor impossible.

Centrifugal pumps have a vast range of applications, starting from pumping water in a residence, from a water sump to an over-head tank; from an open well for irrigation needs of a small farmer to pumping large quantities of water from rivers and canals. Monoblock centrifugal pumps used in open wells have the advantage of being moved to different levels to keep pace with the fluctuating water levels during pre-monsoon and post-monsoon periods. High capacity centrifugal systems pump thousands of liters of water, over high pumping Heads and over long distances as integral components of Lift irrigation projects, and town water supply schemes.

One such major project is the Devadula lift irrigation project, in Warangal District, Andhra Pradesh, the second biggest in the world.

Water is pumped from an intake structure at Gangaram on the banks of Godavari, and is transported in stages, by 2.5 dia steel pipes, over a distance of 135 kms ascending an elevation of 165 m. The pumped waters irrigate 1.6 lakh ha of parched land and provide drinking water to townships and villages enroute.

The pump shown in fig. 18 is one of the 5 pumps which together constitute an integral part of the Project. The pump runs on 8.5 mw power, pumping water at a rate of 10 cumecs (350 cusecs) with a capacity head of 130 meters. The pump forms one of the five large pumps operating at this lift irrigation project.

Fig 18: One of the pumps under erection at Devadula lift irrigation project in Andhra Pradesh

The pump is manufactured by Ebara Company of Japan, at its facility at Haneda. It runs on 8.5 MW of power, and is one of the largest under use in India for pumping water for irrigation. It is installed at the intake at Gangaram on the Godavari river. It pumps 10 cumec (350 cusecs) of water with a capacity pumping Head of 130 meters (425 feet).

The other 4 pumps, of the same capacity are also manufactured by Ebara, each running on 5 MW of power. In tandem, the five pumps lift the river waters against an average pumping Head of 400 meters.

Figure 19 is a depiction of high-power pumps lifting large quantities of water, over high Heads and delivering water for irrigation and water supply projects. The total energy taken by the pump to move water from the source, to the point of discharge is 'total head'. The total head of a powerful turbine pump can exceed several hundred meters.

Fig 19: Schematic illustration of functioning of high-power pumps in Lift irrigation projects adopted from Pomona pumps illustration—courtesy: pump Handbook by Volney C.Finch

Turbine pumps:

A turbine pump is a vertical shaft centrifugal pump that is mainly used to pump water from deep wells or other underground sources to water distribution systems. The system consists of a pump shaft, a rotating device with impellers, and a motor to drive the pump.

The turbine pump motor is placed above the water level either on specially constructed mechanized platforms or on the ground, and connected to the pump, as schematically shown in fig. 20.

The pump may consist of multiple semi-open or enclosed impellers, also known as "stages." A metal shroud supports the vanes of the impeller in an open or semi-open impeller, whereas in an enclosed impeller, the shroud encloses the impeller vanes. Turbine pumps are designed to deliver large volumes of water from deep wells or where the groundwater level is well below the suction limit of regular centrifugal pumps. High capacity turbine pumps are deployed to lift water from the banks of deep river valleys by installing intake wells or tube wells drilled in the vicinity. Turbine pumps, unlike submersible pumps are cost effective, from both the capital outlay and maintenance, as the machine is easily accessible for upkeep and repairs.

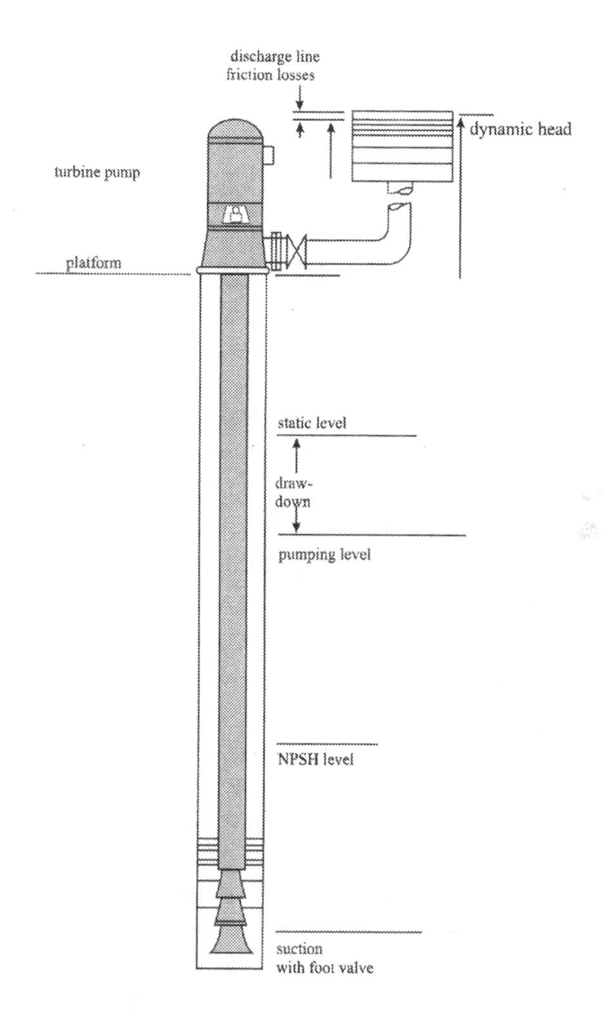

Fig 20: Schematic section of a turbine pump placement

A deep well turbine pump has following major parts:

1. The pump head located at the ground or on a platform.
2. The pump bowl assembly located down in the well, beneath the water surface.
3. The pump also consists of a water intake point and a water discharge point, which is coupled to the discharge column.
4. The discharge column suspends the bowl assembly from the pump head, provides eduction for pumped water, and supports the vertical shaft bearing.

This assemblage is provided with necessary couplings and control valves.

The bowl assembly is made of one or more bowls or stages in series. Number of stages is decided based on the height (Head) to which the water needs to be lifted, calculated from the depth to the pumping water level to the point of discharge plus ancillary Head accruing from friction through pipe lines, valves, and bends. Each bowl contains a rotating impeller which discharges water under high velocity into the diffuser. The diffuser is a stationary ring with vanes arranged to provide passages of gradually enlarging cross sections, and its function is to slow down gradually the velocity of water that it receives from the impeller. Thus, it converts the velocity Head to pressure Head, before the water is led into the suction opening of the next stage. It is the presence of the diffusion vanes that classifies a centrifugal pump as a Turbine pump. The water coursing up the discharge column, after going through the successive stages of the bowl assembly will possess pressure energy equivalent to the combined functioning of all the stages minus the Head losses occurring during its passage from bowl to bowl.

A turbine pump has the following advantages:

1. Installation and maintenance are easy as the prime mover is on the ground
2. Varying Heads caused by seasonal water level fluctuations does not overload the system.
3. The pumping Unit works independently of suction, and remains always submerged, eliminating the need for priming, and other ancillaries required to maintain suction. The pumping action in each of the stages of the turbine pump is similar to the ordinary centrifugal pump. When a number of stages are sequenced in series, the pressure Head produced will be nearly in direct proportion to the number of Stages. Thus if a single stage of the pump will lift 1000 LPM to 10m, eight stages will lift 1000 LPM to 80m.

Submersible pumps:

Amongst all the water well pumps, the most versatile and efficient version of a centrifugal pump designed to pump water from deep wells is the submersible pump. These pumps are widely in use in industrial and agricultural water sourcing installations all over the world. The submersible pumps are multistage centrifugal pumps operating in a vertical position. Although the constructional and operational features of the submersible pumps undergo a continuous evolution because of their versatility and wide-ranging application, their basic operational principle remain the same.

Submersible pumps are commonly is use in many applications. Single stage pumps are in use for drainage, industrial, and slurry pumping. Multiple stage submersible pumps are typically lowered down a bore or a tube well and used for water abstraction from considerable depths. The pump itself is a multi-stage unit with the number of stages being determined by the operating requirements, such as the output, depth from which the water is to be lifted, distance to delivery point from the wellhead and the height to which it needs to be lifted. Each stage consists of a driven impeller and a diffuser which directs flow to the next stage of the pump. Submersible pumps range in diameters from 90mm (3.5 inches) to 254mm (10 inches) and vary between 1 m (3ft) and 8.7 ms (29 ft) in length. The motor used to drive the pump is typically a 440 volts, three phase, squirrel cage induction motor, with a nameplate power rating in the range 7.5 kW to 560kW (at 60 Hz). Most commonly used submersible pumps are those that fit 114 and 165 mm (4.5 and 6.5 inch) diameter wells. Small diameter pumps are available for domestic use. Table 18 shows the nominal diameter of the regular submersible pumps and their discharge and Head capacity.

Table 18: Pump diameter and capacity range

Nominal Diameter (cms)	Discharge (range) (lpm)	Max Head (meters)
10	30 to 120	100
	100 to 300	90
15	50 to 200	190
	200 to 500	160
	300 to 900	150
20	300 to 1400	130
25	600 to 3000	76

The Table is only broadly illustrative; there is a wide spectrum of pumps with a range of 'yield' and 'Head' distinctiveness to suit various hydrological characteristics of any well.

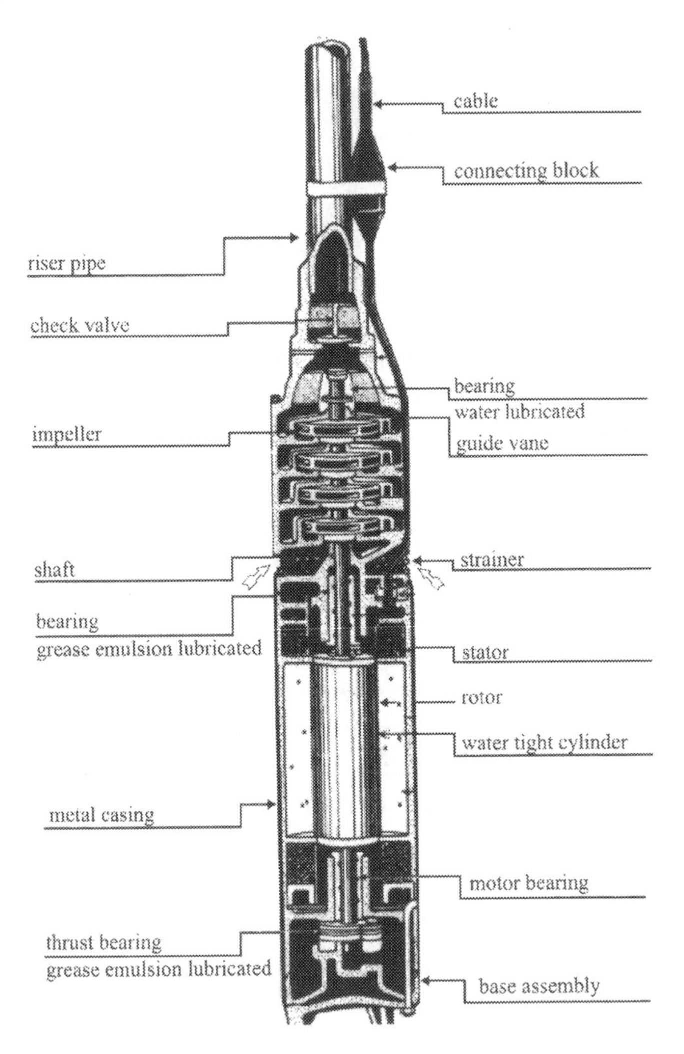

Fig 21: Cutout view of a multi-stage deep well submersible pump adopted from illustration by American Steam pump company: Courtesy: Pump handbook by Volney C.Finch

A submersible deep well pump is a sequenced pumping unit consisting of a turbine pump coupled to a vertical submersible electric motor below it. A mechanical coupling at the bottom of the pump connects the pump shaft. Well water enters the pump through an intake screen and is lifted and moved upwards by the series of stages. Other parts include the radial bearings (bushings) distributed along the length of the shaft providing radial support to the pump shaft turning at high rotational speeds. An optional thrust bearing takes up part of the axial forces arising in the pump, but most of these forces are absorbed by the protector's thrust bearing.

This compact pumping unit is coupled to the end of a discharge column, and the entire assembly is suspended in the well. The discharge column could be of GI piping, or HDPE piping. The later piping makes installation easier, and faster. The Unit operates completely submerged in the water in the well. Hence, it is critical to ensure its continuous submergence by positioning well below the pumping water level, and below its NPSH value. Delivery of water is through the riser pipe coupled to the pumping unit; power is conducted to the motor through, rubber sheathed waterproof cables. All the impellers of the multistage submersible pump are mounted on a single shaft; and hence all rotate at the same speed. Impeller in each stage injects water to the eye of the impeller in the succeeding stage via a diffuser. The diffuser shaped to reduce momentum converts velocity to pressure Head. Each impeller with matching diffuser constitutes a stage; height to which a pump can lift water depends on the number of stages it has. Hence, to match the hydraulics of any individual well, the submersible pump is carefully selected, with emphasis on the number of stages, taking into consideration the depth from which the water is to be lifted, end to end distance over which the water is to be conveyed, and the height to which water needs to be lifted. For pumps pumping in tandem, or pumping in parallel, in addition to number of stages, selection of pumps should also take into account requisite system pressure, and capacity.

Every centrifugal pump manufactured is designed with a specific operating condition, which is an integral arrangement of its motor speed (RPM), Head and yield, which gives the best efficiency for that particular combination. All the three are intricately interrelated. Any change in speed or diameter of the impeller will have a proportionate effect on its efficiency and yield. Head varies as square of the speed of the motor rating and, power required as cube of its speed. Consequently, Yield and Head are inversely proportional to each other for any given pump, which are a function of the impeller diameter and vane angle. Since power required is the product of Head and discharge, same pump is designed with varying Head and discharge factors for machines of similar horsepower. Lower the Head, higher is the yield; and reciprocally, higher the Head lower the output. Since power required is the product of Head and discharge, pumps cannot be rated based on Horsepower alone. Higher horsepower does not automatically imply higher output. The pumps are designed either for low Head and high yield, medium Head and medium yield, or high Head and low yield. There are a range of pumps balancing these two factors viz., yield and Head. This is one of the major reasons for emphasis on proper selection of a pump to suit every individual well. Correct selection is to match Head and yield of the pump with the Head and yield of the well.

For example a 1.5 H.P. pump with impellers designed for high output would deliver more water than a pump with impellers designed for higher Heads, irrespective of the number of impellers. A 1.5 H.P. pump with 7 impellers, designed for yield would deliver more water from 25 m depth than a 5 H.P. pump with 10 impellers designed for high Heads.

Every pump has a performance chart provided by the manufacturer, which indicates the pump's output capacity for various Heads. Reference to the performance charts is an essential prerequisite, prior to selecting the correct pump for any newly drilled well. The pumps are selected based on the pumping test data. However, in the absence of a pumping test, pumps can be provisionally selected from the following data:

i) The volume of water output recorded during drilling, ii) the recorded depths at which water or traces of water appeared during drilling; iii) and the depth to water level measured 72 hours after completion of drilling. These parameters will also constitute the starting point while selecting pumps from diverse performance charts.

Advantages of a submersible pump:

1. The submersible pumps are practically noiseless, and is easily installed. The pumping unit and the motor are coupled; they can be lowered into the well by adding sections of discharge columns until the desired setting is reached. The discharge column could be of either GI pipes screwed one into the other,

or a single, long HDPE pipe, which can be lowered manually without the aid of a tripod. There is no long shaft or bearings to be aligned, and the discharge assembly with the pump can be inserted into wells, which might be eccentric, and vaguely off the vertical, in which it may not be practical to install conventional deep well Units with their long shafts.

2. Number of moving parts is minimum. There is no frictional load accruing from rotating a long shaft at high speeds. The delivery of water to the surface is through an open flexible pipe, instead of around a shaft with intermediate bearings.
3. No pump House is required as the Unit operates completely underground. The control and instrument panel could be installed on an electric pole with a rainproof cover.
4. The Unit is maintenance free over a long number of operating years.

Jet pumps:

The Jet pump is another modified version of centrifugal pump which finds wide application in domestic fluid supply systems. Major characteristic of a Jet pump is its simple design, low cost, quite operation, and least maintenance. A jet pump draws water up from a well, unlike a submersible pump, which pushes water up from a well. Jet pumps have only a suction assembly; while submersible pumps have no suction, but only a discharge assembly. There are shallow well, and deep well Jet pumps; the former are limited to well depths of about 30 m, while deep well jets have no theoretical depth limit. They are successfully in use in wells that are several tens of meters in depth, ranging down to depths exceeding 100 meters.

Jet pumps with horizontal shafts are classified as horizontal pumps, and pumps with vertical shafts as vertical pumps. Similar to turbine pumps, jet pumps also have single or multiple impellers. Pumps with multiple impellers are classified as multi-stage Jet pumps, which possess higher suction and discharge capacities. There are two types of jet assemblies, for each one of the above variations. The most common is the 'Twin type', which comprises a pair of pipes, installed in the well adjacent to each other; one is the discharge pipe which conducts well water to the surface, and the other is the pressure pipe (fig. 22) which jets part of the water back into the well. In a Duplex type Jet pump, the discharge pipe is within the pressure pipe to accommodate both the pipes in a small diameter well.

The Jet pump operates by the combination of two important principles of Physics. One is the centrifugal force that acts outward on a body moving around a center, which results in water being propelled out with force from the periphery of a rotating impeller. All centrifugal pumps operate on this principle, while a Jet pump combines this fact with the Bernoulli principle, which states that the pressure of water in a pipe is inversely proportional to its velocity, viz., pressure decreases as velocity increases. Based on this principle, the ejector in the system is designed with a nozzle and a venturi, as illustrated in fig.22. The nozzle **'a'** functions similar to the nozzle of a garden hose, which creates a high velocity jet. The centrifugal pump on the surface pumps water to the nozzle within the well under pressure, and maintains the combined flow through the system. The water that is being pumped back into the well, via the pressure pipe, exits through the nozzle as a high velocity water jet. This jet, as it exits through the venturi rapidly expands to occupy the full diameter of the eductor pipe. This rapid enlargement of the flow results in a critical drop in pressure. The decline in pressure causes creation of a partial vacuum which draws water from the well, and the drawn water injects into the assembly at point 'b' recovering the original pressure in the system. This water drawn from the well supplements the volume of water circulating through the Assembly, and is discharged through gate valve 'c.'

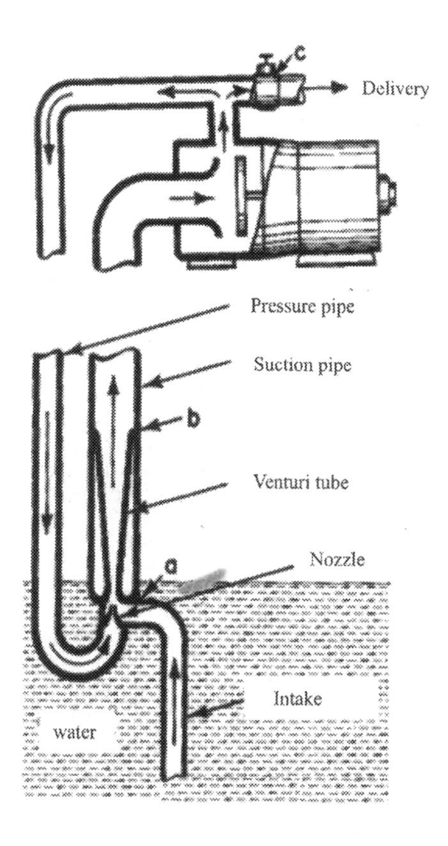

Fig 22: Cross-section showing the functional aspects of a jet pump

The total combined flow discharges past the control valve 'c' which regulates the quantity of flow to the pressure pipe, and to the discharge column. The control valve is regulated to maintain required pressure at the nozzle, and maintain a negative pressure upstream of the venturi to generate flow. Immediately to downstream of the control valve is a pressure gauge to indicate the system pressure. The pump will not discharge any water beyond the control valve, unless enough water is being pumped back into the well through the pressure pipe, to maintain required pressure to force water through the nozzle.

The output of the centrifugal pump varies with any change either in the suction lift, or discharge pressure, ranging between maximum pressure with no flow (means entire output is being pumped back into the well), to maximum flow under nil pressure. System pressure is set on the pressure gauge visually, by coordinating the discharge, and the potential of the well to yield water. Higher the pressure set on the gauge, less will be the discharge, and more will be the water being pumped back into the well. Lesser the pressure, higher the discharge, but the well should have the capacity to sustain the output, without breaking suction. The ejector pumping systems are self-priming. A foot-valve at the terminus of the suction line, below the jet, forms an integral part of the assembly. They are designed to close tightly, without any leakage of water, as soon as the pumping is stopped; and thus maintain the required idle system pressure.

Pumps deployed to draw water from deeper depths, and discharge water to higher elevations need additional number of impellers cased in supplementary stages. Horsepower of the motor to run these additional impellers increase proportionately, but the total volume of water obtained remains constant for any given Head, and varies only with relation to Head.

Advantages of a Jet-pump:

1) Easy to install even in small diameter wells, which may be off-vertical, or under-reamed.
2) Easy servicing of the prime mover, which is installed on the surface. There are no moving parts in the ejector, which stays submerged within the well.
3) Pumps can be offset, and installed away from the well.
4) Suitable for open wells with frequent water level fluctuations. Jet pumps are not sensitive to variation in pumping water levels.

Jet pumps are available over a broad spectrum of Head and Yield demands. Head and Yield being reciprocal to one another, as in submersible pumps, selection of Jet pump to suit a particular well, should take into account the anticipated output (obtained from the Drillers log), and the required Head to which the water is to be pumped. Selection of a suitable pump is made based on this data.

All manufacturers provide 'performance charts, to facilitate proper pump and injector selection. From the charts for each pump, the sizes of the injector nozzle and venturi tube are selected according to the pumping depth, the distance the pump is offset from the well and the size of the pump. Because water is being circulated, involving both suction and delivery in this type of pumping system, it is important that the appropriate sizes of suction and drive lines are used to maximize efficiency. Table 19 indicates the systems correlation for wells deploying jet pumps. The Table indicates, as an example, the typical range of Head and yield relationships of jet pumps, with their respective H.P. / kw rating for various suction lifts, and discharge Heads. From the Table it can be seen that yield reduces as the Head increases; and that the output is not proportional to H.P. of the pump, but is limited by the depth from which the pump is lifting water; pumps with high HP can lift water to greater elevations as indicated in the "Discharge Head Max" column.

Table 19: Correlation between HP/kW, yield, and Head for Jet pumps

For 100 mm diameter wells													
HP/Kw	**Suction lift / depth to pumping water level in meters**											**Discharge head max**	
	9	12	15	18	21	24	27	30	33	36	39	ft	m
	DISCHARGE IN LITRES PER HOUR												
0.5/0.37	2040	1816	1360	900								40	12
0.5/0.37	1360	1360	1250	1135	900							40	12
1.0/0.75	2100	2100	1900	1700	1360	1100	850	680				50	15
1.0/0.75	1350	1350	1250	1135	1000	900	795	680	635	570	450	60	18
	Suction lift / depth to pumping water level in meters												
	10	15	20	25	30	35	40	45	50	55	61	ft	m
	DISCHARGE IN LITRES PER HOUR												
1.5/1.1	2270	2043	1700	1400	1135	953	635					80	24
1.5/1.1	1350	1350	1350	1248	1075	930	726	567	454			90	27
2.0/1.5	2383	2383	2088	1770	1475	1180	953	635				90	27
2.0/1.5	1350	1350	1350	1350	1248	1075	964	830	635	454		100	30

For 115 mm diameter wells													
HP/Kw	Suction lift / depth to pumping water level in meters											Discharge head max	
	10	15	20	25	30	35	40	45	50	55	61	ft	m
	DISCHARGE IN LITRES PER HOUR												
1.5/1.1	2383	2088	1770	1475	1250	1020	795	454				80	24
1.5/1.1	1362	1362	1362	1262	1162	1010	830	635	454			90	27
2.0/1.5	2500	2383	2088	1770	1475	1250	1089	795	454			90	27
2.0/1.5	1362	1362	1362	1362	1362	1262	1162	930	726	635	454	100	30

The pipe sizes vary with the well diameter. Table 20 gives the pipe specifications for the most common diameter wells where jet pumps are installed.

Table 20: Pipe specifications for wells deploying jet pumps

Pipe diameters (mm)			
Well casing	**Delivery pipe**	**Pressure pipe**	**Discharge pipe**
100	30	25	25
115	40	30	25

Reciprocating pumps:

In reciprocating pumps, a mechanical linkage moves a piston back and forth in a cylinder. The piston is driven by a power source, through a crankshaft that converts a rotating action into reciprocating strokes. Water is drawn into the cylinder, through an inlet valve during the suction stroke, and is pumped out through the outlet valve during the compression stroke of the piston. In case the discharge valve is shut, and in the absence of a safety valve, the pressure inside the cylinder would continue to raise, until the discharge pipe or the casing shatters.

The output from a reciprocating pump can be controlled by adjusting the speed with which the prime mover is operating. Discharge pulsates in case of single cylinder pumps, while it is smooth and continuous in machines with twin cylinders. These pumps are made with two cylinders at opposite ends, enabling the pump to draw water into one cylinder, while discharging water from the other cylinder. An air chamber is frequently connected on the discharge flank of the pump to balance the pressure surges. Air in the chamber compresses and expands with each delivery providing an even flow from the well.

Rotary pumps:

Rotary pumps look like centrifugal pumps in their external appearance, but their operation is similar to reciprocating pumps. There are several forms of these pumps. In the simplest expression, two spur gears, or vanes roll together closely, trap the fluid flowing in from the suction, and moves it around to the discharge flank.

When the pump is primed, the fluid will flow in between the gear teeth, as the sprockets rotate, and when it has been carried to the discharge side, there is no scope for it to escape back to the suction. The consequence of these successive deliveries between the gear teeth is the building-up of sufficient pressure to force the fluid through the discharge pipe. Rotary pumps should not be operated against a closed discharge valve.

Air lift pumps:

Water can be pumped by injecting compressed air into the well through a discharge pipe. Air bubbles coalesce with water, reducing the weight of water adequately to bring it to the surface. This is particularly applicable in newly drilled wells, which need further development before they can sustain a regular output. Air-lift pumping is an inefficient method of pumping water, and hence seldom deployed as a permanent installation. High Heads severely affect the efficiency of compressor pumping; hence it is generally employed to lift water to a tank located at the lowest elevation, and a booster installed to lift water to higher elevations. Compact compressors used for air-lift pumping are available in the market; but they should be deployed wherever the yield during drilling has been suspect, and should be replaced with Jet pumps soon after an improvement in yield is noticed.

Hand pumps:

Hand pump is a vertically emplaced reciprocating pump, where the manually operated piston travels up and down in a valve controlled cylinder to draw water from the well. Hand pumps present a simple and versatile means of pumping water from bore wells situated in isolated areas, where power or a mechanical means of extracting water is not available. They are also ideally suited wherever water requirement is not high, or where yield from wells is limited, or have remained undetermined. Hand pumps, if only given the recognition that is rightly their due, have the potential and capacity to resolve Country's rural drinking water difficulties. Insufficient water in many of the rural areas is not as much due to lack of water, as it is to improper utilization of hand pumps, and lack of maintenance of these pumps.

A Hand Pump consists of a metal casing, or Head assembly, installed above ground level, and over the well-head. The cylinder assembly located at the end of the eductor is installed well below the water level, the minimum recommended depth being 8m below the lowest water level in the well. It is made of brass tubing, with the central bore machined to a close tolerance of the plunger rod. The plunger assembly, with an integral valve, and close fitting bucket washers move up and down inside the cylinder. The cylinder contains a non-return valve at its bottom, with a strainer below, to exclude suspended particles in the well water from entering the cylinder. A set of rods generally of 3m length each, coupled with a hexagonal nut provide the linkage between the plunger operating within the cylinder, and the pump handle fixed to the Head assembly on the surface. The eductor column connects the cylinder with the pump fixture on the ground, and conveys water to the surface.

Fig 23: India mark II Hand pump

India Mark II(fig.23) is one of the popular, largest selling hand pumps in the world. It is in use in many countries forming the backbone of the community water supply. It can lift water from a depth of 50 meters. The pump body and its major parts are Hot-Dip Galvanized and as such are extremely durable and long lasting over the years. With little timely maintenance, it can work under strenuous and extreme weather conditions over an extended period.

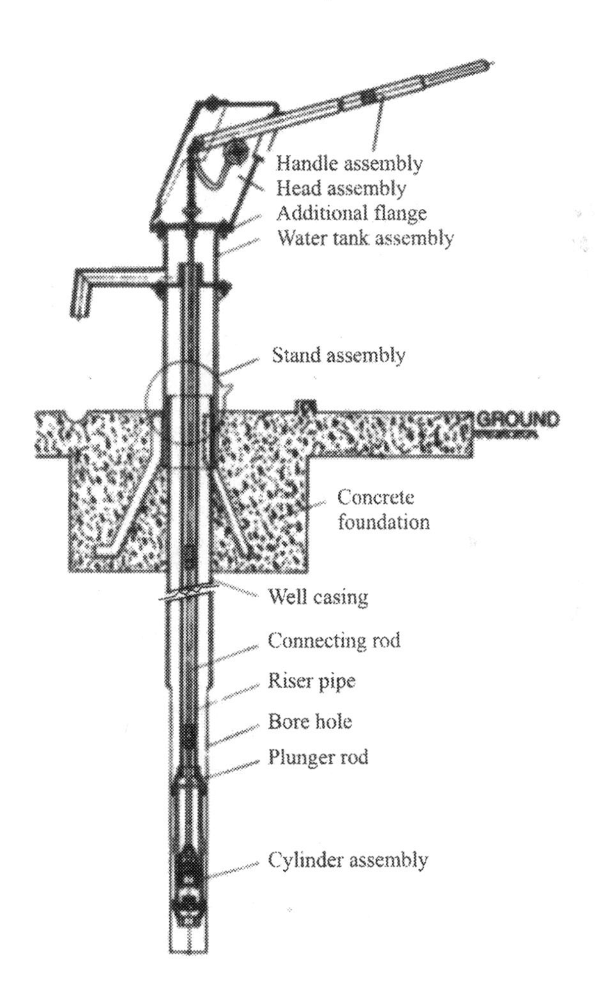

Fig 24: Cross-sectional view of a Hand pump

Fig 24 illustrates the installation and the simple working mechanism of a hand pump. The plunger moves up during the downward stroke of the pumping handle at the surface, and moves down during the upward stroke of the handle. The upward travel of the plunger is the suction stroke, creating a negative pressure in the cylinder, resulting in opening of the lower check valve and drawing water into the cylinder. The downward travel of the plunger is the compression stroke, which automatically shuts the lower non-return check valve in the plunger and opens the valve in the plunger. Water from the lower part of the cylinder enters the upper part through this valve as the plunger continues to move downward. Subsequent upstroke of the plunger raises the water through the eductor pipe until it exits through the pump sprout at the surface. The up and down strokes of the pump handle results in a nearly continuous spout of water discharging from the pump nozzle at the surface.

The main pump consists of three sections, viz., a firm stand for embedding into the foundation, water tank with sprout, and the handle mechanism. The Handle swivel on an axle supported on ball bearings, and is connected to the plunger rod through a heavy-duty chain. A quadrant shaped chain guide fixed to the handle facilitates necessary adjustments, while dampening any lateral vibrations; the entire mechanism being enclosed in a metal box. The mechanism can be fixed on most of the hand pumps with varying designs.

Capacity and limitations:

The capacity of a hand pump is governed by the inner diameter, and length of the cylinder. This capacity is limited, as where high suction lifts are involved, the effort required to operate the handle increases proportionately. The standard design prescribes a cylinder of about 65mm inner diameter, and 300mm length which is appropriate to draw water from a depth of 60 meters. However, there are hand pumps available with specifications to suit diverse needs.

Table 21: Capacities of regular hand pumps

Cylinder inner diameter (mm)	Capacity in LPH @50 strokes/min	Well diameter (mm)	Eductor pipe inner diameter (mm)	Suction limit (meters)
63	1900	102	32-38	60
76	2600	102	32-38	45
89	3600	152	38-51	30
102	4600	152	38-51	25

Cylinder diameters limit the capacity of the pump to lift water. Output capacity of the pump and its suction limits are inversely proportional to each other.

The hand pumps constitute the lifeline and well-being of a vast majority of the villages in many Countries; particularly those Countries that are largely dependent on their rural economy for their progress and growth. India is one such Country, dependent for its advancement on the rural economy. Still, in the vast majority of the villages, many hand pumps remain defunct because of lack of maintenance; denying the residents, much needed water.

Hand pumps, generally installed as common utility, operate under severe conditions, inherent to open unprotected installations. Despite its simple working parts and robust construction, a large number of these pumps remain non-operational, mainly because of rough handling by a huge numbers of users, and partly because of lack of spare parts. The functional life of a hand pump could be prolonged, if periodical attention is paid to adjust or replace the parts that are subjected to working strain.

The parts that may need frequent attention are the pump buckets, which are actually cup shaped washers made of leather or synthetic rubber. Buckets made of very soft leather wear out quickly, but will not damage the cylinder walls, while buckets made of hard synthetic materials will not wear, but may cause scoured cylinder walls. Cylinders are costlier to replace; hence, it is best to use buckets of soft leather, and replace them at regular intervals. Other common causes of breakdown are broken or stuck valves, broken or disconnected plunger rods, and in rare cases a cracked eductor pipe. These are the result of using substandard materials, improper installation, and insufficient maintenance.

Hand pumps installed in wells, which may be off-plumb break down often consequent to excessive stress on bearings and bushes. Fixing nylon guide bushes at suitable intervals on the plunger rods, between two lengths of column pipe facilitate retention of their verticality, and concentricity; minimizing the possibility of breakdowns resulting from installing pumps in improperly drilled wells.

Wells installed with hand pumps may fail to yield water because of the following reasons:

i) insufficient cylinder submergence in wells where the drawdown is rapid;
ii) pumps installed without considering the decline in water levels during dry season;
iii) pumps installed in wells with insufficient casing.

Hand pumps should be selected, from Table 21 or similar pump capacity charts, based on the anticipated yield and depth to water, both of which could be computed from data acquired during drilling. Selecting and installing hand pumps at random invariably leads to their failure in course of time; and is at present the major cause for hand pumps installed in remote, isolated rural areas malfunctioning leading to deficiency of water.

Proper drilling, careful installation, maintenance, gentle handling, and stocking of the important tools and spares, in addition to training some local persons about the pump maintenance would go a long way in keeping these hardy pumps functional, assuring year round water supply in isolated tracts.

CHAPTER XI

SYSTEM VALVES

Valves are essentially restricting devices, which form an essential and integral part of the pumping and transmitting systems. They perform major functions in regulating either the flow, or the pressure of the flow from the point of origin, through pipelines, to the point of usage. They ensure a smooth one-way flow through the system, by preventing back flow, and build-up of sudden and undue backpressure on the pump. However, when water flows through a valve, whether fully or partially open, it loses some energy, which results, depending on the flow rate, in accrual of additional Head on the pump.

Main valves necessary for the majority of the water pumping systems are the Check valve (foot valve), Gate valve and the Air-valves. Check valve, also known as clack valve or non-return valve allows water to flow only in one direction, and automatically prevents reverse flow and thus ensures non-occurrence of excess static load on the pump. There are various types of check valves. Check valves work automatically and do not require any manual control. Most do not have any valve handle or stem. The bodies (external shells) of most check valves are made of plastic or metal. An important variation in check valves is where the valves are spring-loaded to specified cracking pressure which is the minimum upstream pressure at which the valve will open. Typically, the check valve is designed for specific cracking pressure.

Most common are the swing check valves'(fig.26) or 'tilting disc check valves' in which the disc, the movable part which blocks the flow, swings on a hinge either onto its seating to block reverse flow or off the seating to allow normal flow. The seat-opening cross-section may be perpendicular to the centerline between the two ports or at a vertical angle to the pipeline. Its typical location is downstream of a pump as shown in fig.28 and at the terminus of an eductor in a bore-well. Placed at the end of the pipeline inside a well, it ensures a standing column of water in the pipeline allowing automatic priming of the pump.

Multiple check valves can be connected in series. A double check valve is frequently installed to prevent potentially contaminated water from siphoning back into Town water. There are also double ball check valves in which there are two ball and seat combinations sequentially in the same body to ensure positive leak-tight shutoff when blocking reverse flow; and piston check valves, wafer check valves, and ball-and-cone check valves.

Foot valves (fig 27) are non-return valves installed at the end of the pipeline inserted into the well. All centrifugal and suction pumps will deliver water only when the entire pumping assembly, consisting of the pump casing and the piping are devoid of any air pockets, and are well submerged in the water supply source. Bringing the system to this state is called 'priming' and most of the pumps are self-priming only when the foot valve at the end of the line blocks water in the system from migrating back into the source when the pump is idle. These valves control back flow, when the pump is shut down, as they automatically close to prevent water in the pipeline from returning into the well. These valves passively maintain system pressure, retaining water in the idle pump, consequent to which the pump will remain in the primed condition, facilitating instant discharge on starting the pump. The area through the foot valve should be one and half times the opening of the suction pipe, and should be of the type where a flap valve swings up and out of the way, while water is being drawn into the system. Foot valves should not be deployed, when a centrifugal pump is running against a high dynamic Head, unless a reflux valve (check valve) is installed in the discharge line. In the absence of a check valve, an accidental power stoppage would result in water in the line rushing back through the pump to impact against the shut foot valve, causing destructive hammering action on the impellers. Strainers cover foot valve inlets, but wherever the well is unclean with flotsam, a larger sized secondary screen is wrapped around the entire suction intake, to prevent gradual choking of the strainer. Such a secondary screen should have openings large enough to limit the entrance velocity to 60 cms/second.

Friction developing in the flow during its passage through foot valves should be at its minimum. Friction accruing in the valve is computed by the simple equation:

$$Hf = KV^2 / 2g$$

Where,

Hf = friction loss;
V = Velocity of flow in the suction pipe and
K = friction factor of pipe
g = 9.75 m

Frictionless foot valves have a friction factor 'K' less than 0.8 and 'K' is less than 0.5 for reflux valves. Reflux valves are deployed in pipe lines traversing a gradient.

Standard foot valves, designed and manufactured as per IS:10805 are friction less foot valves. These swing type non-return valves are screwed into the suction pipe, while the reflux type friction free foot valves have screwable threads at both ends. These valves, with a large body facilitate a complete and maximum swing for the valve and ensure a straight, unrestricted, and smooth flow for the water being drawn into the well. This reduces load on the prime mover, saves energy to a considerable extent, and affords optimum discharge. Reflux type non-return valves are deployed in the delivery pipe-line to prevent water hammer, and backspin effects on the pumping system. It has been estimated that use of friction less foot valves increase the discharge of water by nearly 20%, while reducing power consumption by about 10%.

Gate valves (fig. 25) are fitted along the conduit system to shut-off flow, while throttle valves regulate flow. Gate valves are not suitable for controlling flow, as they will pass maximum flow even in partially closed conditions. They are deployed for complete shut off of the system. For throttling or regulating flow a globe valve is deployed. Installed at the end of a discharge pipe of a pumping well, downstream of the pressure gauge, they are particularly valuable in preventing over pumping a well. The valve operates by gradually lifting or lowering a metallic disc, a rectangular gate, or a wedge across the flow path. Gate valves are primarily used to allow or shut-off the flow of water, and typical gate valves should not be used for regulating flow, unless they are specifically designed for that purpose.

Fig 25: A gate valve

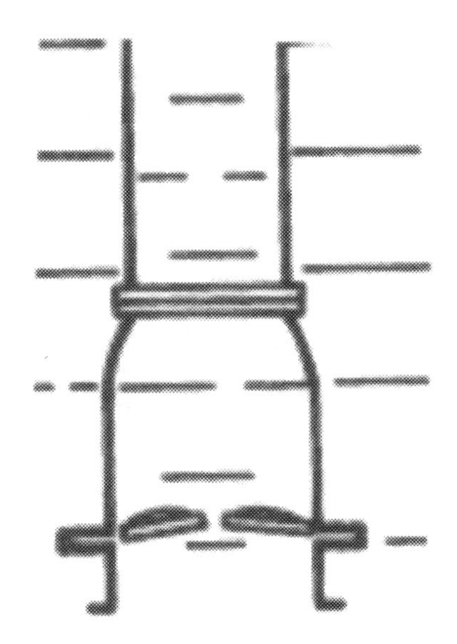

Fig 26: A foot valve

Fig 27: Frictionless foot and reflux valves

Gate valves have either a rising or a non-rising stem. Rising stems provide a visual indication of valve position because the stem is attached to the gate, and the gate and the stem rise and lower together as the valve is operated. Non-rising stem valves may have a pointer threaded onto the upper end of the stem to indicate valve position, It is often critical to monitor the amount of opening of the gate, such as during a pump test, and while calibrating the volume of flow with a pressure gauge and the opening in the gate valve.

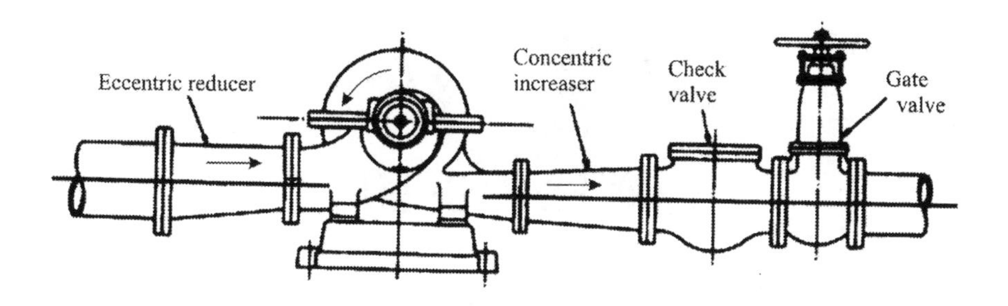

Fig 28: Correct positions for check and gate valves in a conduit system (original system depiction by Dayton-Dowd company. Courtesy: Pump Handbook by Volney C.Finch)

Gate valve is located downstream of pump to control flow (fig 28). Check valve located between the pump and the gate valve will prevent back flow from the inlet pipes to the pump, and protect the system against undue backpressure. The Gate valve shuts off discharge; the valve is closed completely while priming the pump, during starting operations and following shutting down the pump.

Air valves are provided in the pump casing to facilitate any enslaved air (or gas) bubbles to escape; or wherever there is a tendency for the air to detach while traveling through the pump passages. Air is drawn into the suction, consequent to whirlpool action, if the inlet submergence is less than the recommended depth, or empirically less than 1 meter. Air leaks in the suction are another source of air being drawn into the flow. Such leaks should be detected and closed with sealing compounds. A one percent (by volume) air leak can cause the pump efficiency to drop by as much as ten percent, and this volume of air leaking into the system may cause the pump to lose its prime, and cease functioning.

CHAPTER XII

PUMP SELECTION

A pump does not develop power of its own; it needs a diesel, or power energized prime mover to make it function. When a pump is mentioned, it generally implies a coupled unit of a prime mover and a pump. There are multiplicities of pumps and differing pumping systems. The pump to suit each well requires careful selection.

It is incorrect procedure to select a pump at random. The pump (whether a submersible, jet, or turbine pump) has to be cautiously selected to be appropriate to the hydraulic characteristics of the well. Every well whether drilled or excavated, has certain unique hydraulic characteristics of its own, particularly regarding its inherent pressure Head, and its capacity to yield water. Head implies, the pressure with which water will rise in the well above the water table, and yield is the volume of natural inflow of water into the well. Pump selection inherently implies choosing a pump whose pumping characteristics match the hydraulic distinctiveness of the well. A pump so selected and installed will discharge the optimum quantities of water the well is capable of yielding. Pumping a volume of water within the capacity of the well will also essentially ensure that the well does not unexpectedly fail.

Frequently, the well and its location is held responsible for insufficient yield and recurring high power consumption, or pump malfunction, while the actual error stems from improper pump selection. Additional effort and cost expended towards accurate pump selection invariably pays for itself, by the correct volume water that it would abstract from the well, and the lower power costs.

Wells installed with pumps calibrated to the output potential and the pressure Head of the well have yielded nearly three times the water from a well drilled to same depth in the same location fitted with an expensive, high horse power machine selected without taking into consideration the well hydraulics.

The selected 'Head' of the pump should include the height to which water from the well needs to be lifted from its pumping level in the well to the height of the discharge point above the ground level. Pumping level is the depth in meters to the lowest water level when the well is discharging water during dry season—when the water levels will be at their lowest.

A pump's vertical discharge "pressure-head" is the vertical lift in height—usually measured in feet or meters of water—at the end of which, the pump can no longer exert enough pressure to move water. At this point, the pump may be said to have reached its "shut-off" head. In the flow curve chart for a pump the "shut—off head" is the point on the graph where the flow rate is zero.

If the discharge of a centrifugal pump is pointed straight up into the air, water will pump to a certain height—or head—called the 'shut off head'. This maximum head is mainly determined by the outside diameter of the pump's impeller and the speed of the rotating shaft. The head will change, as the capacity of the pump changes.

Selection of an appropriate pump for the well takes into account primarily the 'Head' and the 'yield' of the well. 'Head' consists of many parameters which together constitute the 'Total dynamic Head'.

Total Dynamic Head:

Total Dynamic Head or the TDH, is the pressure Head that would eventually accrue, and impinge on the pump. It is a critical factor governing pump selection. This important factor is computed from:

1) 'Static discharge Head': This is the difference in elevation between the point of discharge (viz., over—head tank) and the static water level in the well. (Static water level is water level in the well when it is not being pumped)
2) 'Dynamic Head': While the pump is started, and as the pumping progresses, this Head continuously increases as the water level in the well declines, until the well attains equilibrium conditions. At this point, pump is pumping to its maximum capacity lifting water to the highest required

location. This level difference plus the friction accruing from the pipes and fixtures is the 'Dynamic Head'.

3) 'Pumping level' is pre-determined by a pumping test; if such a test has not been conducted, then the probable pumping level should be estimated by adding an empirical figure (about 30 m in hard rock areas) to the static water level. This unempirical selection generally results in high power consumption, without commensurate water yield; but is better than placing the pump at a random elevation.
4) Total length of the pipe lines conducting water to the point of usage.
5) Type and number of fixtures in the pipeline.
6) Diameter and nature of the conducting system (pipe lines).

As water flows along the pipelines, it loses certain amount of pressure, or Head, because of friction, and this is called 'Head loss' due to friction. This is added to the suction and discharge Heads to obtain TDH.

Diameter of the pipelines has a significant bearing on the TDH acting on the pump. Larger the diameter lesser the frictional Head acting on the system, and lesser the required HP with real-time reduced power consumption. Large diameter pipelines need higher capital outlay.

TDH is expressed by the equation:

$$TDH = hv + hf + hc$$

Where:

TDH = total dynamic Head in meters;
hv = total vertical lift, from low water pumping level in the well to the highest point of delivery, in meters;
hf = total friction losses accruing in the conduits, expressed as Head, in meters;
hc = velocity Head in meters = Head required to produce flow.

When the pump works against system pressure (Head sustained in the system as a result of auxiliary pumps pumping in parallel), rather than simply lifting water to an overhead tank, total lift 'ht' is computed from the vertical distance from the pumping level in the well to the pump discharge pipe plus the hydrostatic pressure Head at that point.

Velocity Head is obtained from the equation:

$hc = 0.0155\ V^2$. Velocity Head needs to be taken into account only where volume of discharge is very high with a low Head, or where Velocity (V) of flow is very high. If velocity is 2 m/sec, velocity Head is about 1.8 cms.

TDH and Discharge:

There exists an inverse relationship between TDH and discharge. The pumping rate 'Q' decreases with increasing TDH, if the RPM is constant. For this reason impeller type pumps are also called variable discharge pumps. At any given speed, Q varies proportionately with the width, and diameter of the impeller, and TDH increases with the square of the impeller diameter.

Computing well yield and Head:

A simple yield test, as distinct from an Aquifer performance test should precede pump selection. Data from a yield test supplements the well output figures obtained during drilling. Maximum discharge measured with a 60° or 90° V-notch forms the basis to further project the potential of the well. After completion of drilling, the well is developed by pumping with the compressor for about 2 hours until the V-notch records a maximum. This yield in GPH or LPM represents the 'yield' during drilling.

'Head' is computed from the pumping water levels in the well. Pumping water levels become available only during a well test. In the absence of a well test, pumping level is projected by deducting about 12 meters from the static level, which can be measured 48 hours after completing the drilling. As an alternative, during drilling, the water level in the drill hole could be noted at intervals of about one hour. When the drill rods stand disconnected, for adding drill rods to the string, water level measurements are taken through the drill rod, with either an indicator or a chalked tape. These unorthodox methods afford a very approximate level to which the water levels may drop during pumping, but provide a starting point for computing the 'Head' required for the pump purchase. This method is not valid for rotary drilled holes since drilling mud present in the well will give misleading data. In case a pumping or well test is not conducted, these measurements will help in estimating the pressure 'head' of the well.

Approximately 15 m is added to the recorded lowest level to proximate the levels to which water levels may finally drop during actual pumping. These two values viz., the discharge and the computed pumping level would form the basis for selecting a pump whose pumping capacity, and the Head would match that of the well.

Selection of appropriate pump:

Submersible and jet pumps could be broadly classified as "High Head-low Yield"; "Medium yield-Medium Head"; "Low Head-High Yield" pumps. Yield and Head have an inverse relationship; and hence pumps designed to pump large quantities of water cannot lift water from a great depth, and pumps designed to lift water from great depths cannot pump a large quantity of water. Yield and Head of the Pump selected, as far as practicable, should match the yield and Head of the well. Mismatch of these two critical characteristics will result in less water being drawn than what the well is capable of yielding, and will cause high power consumption. Wherever, the installed pump has yield and Head characteristics higher than the inherent capacity of the well, the well will suffer severe degradation, loosing water levels during summer, and eventually will cease to function effectively.

HP of the pump is not significant. It is a frequent misconception that high Horse power pumps can yield large quantities of water. For example, a High HP pump will not extract much water even if the well has a high potential, unless the Head and yield capacity of the pump are coordinated with the Head and yield of the well. A study of Table 23 will show the importance of a careful selection of the pump for any individual well, and that the output does not depend only on the HP of the pump.

Table 22 is an excerpt from the maximum and minimum performance rating of some of the submersible pumps, to illustrate the relationship between the Head and Yield of these pumps. First column signifies a distinctive model Number, while the other columns summarize the capacity of that pump. Last three columns show the output in cumecs and liters per second and total Head respectively. The Table manifests the inverse relationship between the output and Head;

and the reciprocal relationship between the HP, the number of stages, and the output. Curves plotted on graph sheets showing the pump characteristics also indicate the inverse relationship between the yield and the Head, and the relationship between the Head, the output, HP and the efficiency of the respective pump

Table 22: Relationship between Horse Power, Stages, Output, and Head in submersible pumps

Model	Well diameter	HP	kW	Stages	Outlet pipe (mm)	Output (cu. meters/hr)	LPS	Total head (meters)
R 30	80 mm (3")	1.5	1.1	22	25	0.72	0.2	93
R 33	80 mm (3")	1.5	1.1	15	32	4.32	1.2	33
S4G	100 mm (4")	7.5	5.5	14	65	25.2	7	33
S4D	100 mm (4")	6	4.5	60	32	0.72	0.2	400
SM65	150 mm (6")	15	11	5	100	79.2	22	22
SR65	150 mm (6")	15	11	30	50	5.4	1.2	276
CM74	175 mm (7")	15	11	4	100	32.4	9	58.5
CM75	175 mm (7")	20	15	4	100	97.2	27	39
CR82	200 mm (8")	40	30	9	80	56.98	15.83	120
CR82	200 mm (8")	35	26	11	80	18.00	5	214

Hence, before selecting and installing a permanent pump, it is crucial to test the well either with a test pump or with a compressor. Actual yield of a well, with a pump is generally 40 to 50 percent more than the output recorded during drilling, in soft rocks; and 30 to 40 percent more than what is recorded during drilling, in hard rocks.

Tables23 to 26 are extracts, abstracted here as examples, from performance charts (provided by the manufacturer) of a range of submersible pumps designed for installation in 100mm, 150mm, 200mm and 250mm diameter wells. There is a range of pumps to suit the hydraulic situation of any well. The study of the performance charts of the pumps is important as they facilitate selection of appropriate pump to suit the well, and its proper placement elevation.

It is also proper to call these charts as "pump selection charts" applicable to any individual well. The selected pump should match the data obtained during drilling of the well. Essential data is: i) discharge measured during drilling (by the v-notch); ii) levels or depth from which the well has been yielding water; and iii) the water level in the well measured 72 hrs after completion of drilling. To this level, another 40m (120 ft) is normally added, to account for depth of drawdown while pumping. In a typical performance chart, the first and second columns show respectively the horse power (HP) and the number of stages (Stgs) of the submersible pump required, to draw specific quantities of water, shown in the top row. The columns below the respective 'yield' indicate the 'Head' or level in the well, in meters, from which the water is to be lifted. Volume or yield of water, depending on its quantum is indicated in liters per minute, liters per second, or sometimes wherever the volume output is large, in cumecs (cu m per second).

High capacity 75 HP submersible pumps can discharge up to 3900 LPM (860 GPM), lifting water over a Head of 49 m, or 1200 LPM from a depth of 91 m. There are still larger, special purpose wells, which need turbine pumps to abstract large quantities of water.

Submersible pumps are selected, based on the projected Total Head (lowest level from which the pump is lifting water plus the level or height to the point of usage) and the water output (+30%) measured during drilling. Pump selection or performance charts show the capacity of various pumps in terms of Head and discharge. As an example, following is an extract from (courtesy CRI pumps) such a chart which shows the discharge capacity vs Head of various pumps. Top row shows the yield in liters per second; lower rows indicate the Head from which this quantum of water can be lifted and delivered. In order to obtain the maximum quantity of water from the well, and to achieve the objective of drilling the well, it is of critical importance that the performance charts of the pumps are studied carefully to understand the concept of 'Head', and the volume of water it can pump at that Head.

Table 23: Performance chart of a submersible pump for a 100 mm well *(courtesy cri pumps)*

			Outlet pipe diameter = 40 mm									
HP	**Stages**	**LPS >>**	**0**	**0.9**	**1**	**1.2**	**1.4**	**1.55**	**1.8**	**2**	**2.4**	**2.8**
1.5	9	Head In Meters	66	58	56	52	46	39	32	22		
2	12		88	76	73	68	59	50	40	28		
3	15		111	95	91	84	74	62	49	34		
3	18		131	110	107	99	89	76	60	42		
5	25		180	152	147	135	121	103	85	60		
5	27		192	164	158	148	133	113	92	68		
5	32		232	199	193	178	159	138	112	81		
1.5	7		51			45	43	42	38	37	34	27
2	9		66			58	56	54	49	47	43	35
3	14		105			90	86	83	74	72	66	53
5	23		169			150	143	137	124			

Reading the chart: The first and second columns show respectively the horse power (HP) and the number of stages (Stgs) of the submersible pump required to draw 'liters per second' (LPS) of water shown in the top row. The columns below the respective 'yield' indicate the 'Head' or level in the well, in meters, from which the water is to be lifted. For example, the Chart indicates that a 1.5 HP pump with 9 stages can lift and deliver 1 lps (800 imp gallons per hour) water from a max depth of 56 m; OR 2 lps from a depth of 22 m. To draw 1 lps water from a depth of 193 m, a 5 HP pump with 32 stages would be required.

Table 24: Performance chart of a submersible pump for a 150 mm well

Outer diameter of pump = 143 mm, Outlet pipe diameter = 50 mm									
HP/kw	**Stages**	**GPM>>**	**0**	**11**	**14**	**18**	**22**	**26**	**33**
		LPM>>	**0**	**0.83**	**1.08**	**1.33**	**1.66**	**2.00**	**2.5**
2.0/1.5	5	Head In Meters	48	44	42	39	35	30	20
2.0/1.5	6		58	53	51	48	43	36	24
3.0/2.2	8		77	71	67	63	59	49	32
4.0/3.0	10		96	88	84	79	71	61	41
4.0/3.0	12		115	106	101	95	85	73	49
5.0/3.7	14		134	124	118	111	99	85	57
6.0/4.5	16		154	141	135	127	114	97	65
6.0/4.5	18		173	159	151	142	128	109	73
7.5/5.5	20		192	177	168	158	142	121	81
7.5/5.5	22		211	195	185	174	156	134	89
9.0/6.7	24		230	212	202	190	170	146	97
9.0/6.7	26		250	230	219	206	185	158	105
9.0/6.7	28		269	248	236	222	199	170	114
10 /7.5	30		288	265	252	237	213	182	122

The chart indicates that a 2 HP pump with 5 stages can be deployed only if the 'Head' (pumping depth to water + the height to which water is to be pumped) is 44 m or less. At 44 m it can pump 11 GPM; and if the 'Head' is only 20 m, the same machine can pump 33 GPM. A 10 HP pump with 30 stages can yield 11 GPM if the Head involved is 265 m; and 33 GPM if the 'Head' is 122 m. In the middle are a range of pumps with high yields for low Head requirements, and low output for wells with a high Head.

Chart indicates output in liters per second against different

'Heads'. Columns 1 and 2 represent ten different pumps to match respective 'Heads' and 'output'. For example a 20 HP pump with 4 stages can yield 18000 liters per hour lifting water from a depth of 76 m or 87000 liters per hour lifting water from a depth of 34 m.

Table 25: Performance chart of a submersible pump for a 200 mm well

Outer diameter of pipe = 195 mm, Outlet pipe diameter = 100 mm												
HP	Stages	cu.meters/hr>>	0	18	24	39	50	63	69	75	81	87
		LPS>>	0	5	7	11	14	17	19	21	22	24
20	2	Head In Meters	49	48	47	46	45	43	42	41	40	38
30	3		83	78	77	72	68	64	63	61	60	57
35	4		106	102	101	97	94	88	82	81	77	74
40	4		111	106	104	99	97	91	89	86	83	80
10	3		50	48	47	42	38	35				
10	2		48	42	40	38	36	34	33	32	30	28
15	3		64	61	60	56	54	48	46	44	42	39
20	4		80	76	75	72	65	60	59	53	47	34
40	8		160	149	148	136	126	108	100	87	72	60
50	10		205	188	184	170	160	146	130	120	100	82

(*courtesy: CRI pumps, Coimbatore*)

Table 26: Performance chart of a submersible pump for 250 mm

Outer diameter of pump = 247 mm, Outlet pipe diameter = 150 mm											
HP	**Stages**	**cu.meters/hr>>**	**0**	**18**	**36**	**54**	**72**	**90**	**108**	**126**	**144**
		LPS>>	**0**	**5**	**10**	**15**	**20**	**25**	**30**	**35**	**40**
40	2	Head In Meters	70	69	68	66	64	62	60	58	54
60	3		102	100	97	95	93	90	87	83	79
100	5		164	161	157	154	150	145	139	132	125

Characteristic curves of a pump:

Every pump has a characteristic curve. Characteristic curve is the graphical representation of the correlation between out-put, Head, H.P. and efficiency. Important operating features of a pump are its discharge capacity and its Head at a constant R.P.M, and hence the principal characteristic curve represents the relationship between the two. Graphs displaying this feature show discharge in LPM on the abscissas, and Head in meters on the ordinates. Discharge varies with Head. As the throttle valve is closed, out-put reduces, and Head increases proportionately. The Head-Output curve slopes downward from left to right, continuously reflecting the increase in discharge, as Head declines. Zero delivery point indicates the "shut-off" Head when the throttle valve is fully closed.

The characteristic may be steep, or flat, depending on the impeller design, but in all cases would be continuously descending, from the point of zero discharge to the point of max discharge. As the curve descends to max, the velocity in the impeller eye will become so high, that the curve will abruptly drop vertically. This is the "cut-off" point of the pump.

Most of the pumps plot a stable curve, where there is only one rate of discharge for any specific Head, the discharge continuously dropping as the Head increases. Pumps plotting a steep curve reflect capacity to maintain discharge over a small range even if there occurs a significant variation in Head. Unstable curves indicate a pump with a peak 'Discharge-Head' single point, where the out-put is maximum for that specific Head, beyond which the discharge will decrease even if the Head is lowered. In a drooping curve, the plot initially rises rapidly from 'shut-off', with both Discharge and Head increasing concurrently over a small range, and then descending steeply. Relatively flat curves mean that small changes in Head result in considerable variations in out-put.

Ideal procedure of proper pump selection is to test the well for its yield and Head, plot the data on a tracing paper in the same scale as that of the pump characteristic curve, and match both the curves by an over-lay, and select the pump whose curve matches or nearly matches the well curve. If plotting curves appears impractical, second alternative which is less precise, is to study the tabular selection charts of a number of pumps, and select a pump whose Head and yield matches, or nearly matches the well test data.

NPSH, NRIH, TRIH Coefficients:

Centrifugal pumps, whether surface or submersible, operate by creating a low pressure at the inlet or the intake portal, which allows the water to be pushed into the pump. This force acting at the intake is the pressure exerted by the standing column of water plus the atmospheric pressure, both of which are variable. There is a physical limit to the pump's operation, based on these external factors. This limit is the Net Positive Suction Head (NPSH), which is the absolute pressure available at the pump suction. It signifies the energy Head required to overcome pump's internal Head losses, accruing as water ingresses, and accelerates through the inlets and impeller vanes of each stage. In the pump characteristics, NPSH curve relates the suction head required by the pump to its output. It varies with the make of the pump, with different models of the same make, and continuously changes with the output, and running speed of even one pump. NPSH needs to increase with higher discharge.

For a centrifugal pump, the term 'inlet' is the eye of the lowest impeller. The term 'intake' denotes the entry point to the piping system. For efficient pump operation, the system should be under sufficient pressure Head to equal, or exceed the NPSH. Every pump, in addition to the system NPSH, requires a certain minimum pressure at its inlet to keep the pump running efficiently, and without causing cavitation.

This pressure, specified in terms of Head of water, either in meters or feet is NRIH—Net Required Inlet Head. NRIH is a function of the pump, and is either specified by the manufacturer either in the pump manual or inscribed on the pump casing. NRIH can be more, or less than the NPSH. The available pressure at the pump inlet—the NPSH—can be computed from the functioning system, while the NRIH is provided by the manufacturer. NPSH must be maintained at levels higher than NRIH for the pump to run efficiently. It implies that pressure at suction should be greater than that required by the pump. If the pump is allowed to operate at the available NPSH than that required by the pump, many incipient but serious, cascading problems accrue in course of time. Hence, it is critical to obtain the

NRIH of the pump from the manufacturer, and to ensure that NPSH pressure in the system will be adequate to satisfy the NRIH requirement.

Water, like most liquids transform from a liquid to vapor state, depending on ambient temperature, and pressure. Liquids with high vapor pressure change into vapor state rapidly. Some remain in liquid state only at pressures exceeding atmospheric pressure. There are liquids with such low vapor pressure, which will not readily change to a vapor state even under vacuum. Water has a median vapor pressure, which increases rapidly with rise in temperature.

Vapor pressure is a critical factor in the operation of centrifugal pumps, as it limits suction lift. During pumping, if the pressure at any point in the suction line decreases to vapor pressure of water, air bubbles form and are entrained in the water. As water moves, these bubbles collapse violently—called implosion—when they reach points of higher pressure as they transit through the system. When the flow enters high-pressure areas, bubbles implode or rapidly collapse on metal surfaces causing pitting and metal fatigue. The shock waves produced during implosion or collapse, are strong enough to damage moving parts. The pump impellers and bends sustain most of the damage. This implosion results in pitting in pump parts, and is called 'cavitations'. Cavitations interfere with pumping, causing excessive vibration. Cavitation occurs when the system continuously operates, below NPSH value, under low hydraulic Head over the pump intakes. The Head should be sufficiently high at the point of entry into the intake, so that as the velocity of water increases, the concurrent pressure drop does not go below the vapor pressure.

This requires a total Head at the pump inlet equal to, or more than the specified NPSH (absolute) of the pump, plus NRIH, which includes vapor pressure of water, plus entrance and friction losses in the system. This value is called as TRIH (Total required inlet Head). For any particular installation, total available inlet Head must not be less than the pump's TRIH. In case the pumping level declines close to the pump inlet, total inlet Head available, at any point of time is the atmospheric pressure (expressed in meters of water column), minus the vertical distance from the pumping level to inlet. If the pumping level stabilizes above the pump inlet, available inlet Head is atmospheric pressure plus the vertical distance from the pumping level to pump inlet. If TRIH is greater than atmospheric pressure, the difference between the two is the minimum depth to which the pump inlet should be submerged below the pumping water level in the well. The TAIH (Total available inlet Head) can be increased by lowering the pump intake to a lower level, or decreasing the drawdown by means of partially closing the throttle valve, thus maintaining a higher pumping level.

In case of surface centrifugal installations where pumping level is lower than the pump inlet, total available inlet head is absolute atmospheric pressure.

Water Horse Power and Brake Horse Power:

The useful work done by a pump is the product of the pumping rate and the total dynamic head against which it is lifting water. The power needed to operate a pump at the desired values of Q and TDH is the 'Brake HorsePower (bhp)'.This is the energy input required at the drive shaft. The power output by the pump is to some extent less than the input at the drive shaft, consequent to losses in pump efficiency.

The power output by the pump in terms of lifting water over a certain height, and moving it over a certain distance is called the 'water horsepower' (whp).

Water Horse Power is computed as follows:

$$whp = \frac{Q \times TDH}{6480}$$

Q = Discharge in m^3 / day
TDH = Total dynamic head in meters of water
1 HP = 746 watts (0.746 kw)

Consequent to friction, back slippage of water, and other energy losses inside a pump, whp is always less than the bhp. The ratio between the whp and bhp is the pump efficiency 'e'

$$e = \frac{whp}{bhp}$$

is expressed as a percentage of the pump's rated HP.

For a specific impeller and operating speed, 'e' is not constant, but varies as Q and TDH change. Maximum 'e' is achieved at only one exacting combination of Q and TDH, which is a function of the impeller and volute design. As Q increases, bhp rises, attains a peak, and then declines as TDH commences to fall.

Brake Horse Power is computed by:

$$bhp = \frac{Q \times THD}{75 \times e}$$

Where:

e = Pump efficiency expressed as a fraction.
Q = Discharge in liters per second.
TDH = Total head in meters.

Centrifugal pumps, run by diesel motors, should be so selected as to have max efficiency at the operating Head.

Following minimum pump efficiency should be ensured:

Pump BHP	Minimum required efficiency
Up to 2	50%
2 to 4	55%
4 to	60%

For centrifugal pumps operated by electric motors, bhp of the electric motor should be 30 to 50% more than the bhp of the pump.

Selection of pumps for Agricultural application:

Pump is the centrifugally operating unit that actually impels water through the pipes; while the unit that drives the pump is the prime mover, which could be either a diesel or electric motor.

Most of the small and medium sized agricultural holdings require diesel or power driven, surface mounted, shallow well centrifugal pumps. Larger agricultural tracts and farms need deep well turbine or submersible centrifugal pumps.

Surveys conducted in India, by various Government and financial Agencies have brought out that in a very large number of cases:

1) pumps fitted on wells do not correspond to the yield of the well and Head,
2) are wrongly installed,
3) have mismatched pump and motor, and
4) incompatible suction and delivery pipe lines.

Deployment of such incompatible pumping systems results in under utilization of the well, besides unproductive expenditure on energy, and maintenance. This faulty procedure spread over a large number of wells in the Country has resulted in wastage of massive quantities of water and power. However, no method has yet been evolved to educate the common user, particularly in the rural areas about the critical importance of selecting and proper installation of a pump. Until and unless this lacunae is set right, any amount of power generation will not meet the power requirements of the rural economy.

Selection of diesel engines as prime movers:

Diesel engines are classified as *slow speed, medium speed, and high speed* machines with their RPM's ranging from 650 to 3600 respectively. For agricultural pumps lifting water from shallow wells within centrifugal limits, power required from the prime mover range between 3 and 15 HP at an RPM rating of 1500. Diesel engines have the following two broad ranges of RPM:

650 to 1800 RPM
1800 to 3600 RPM

Diesel consumption is a function of HP developed, and RPM. A 5 HP machine develops a HP of 3 at 1000 RPM, and 6.5 HP at 2000 RPM. Fuel consumption is high at both extremes, and is minimum between 1250 and 1750 RPM, when it develops the rated 5 HP; and the pump attains maximum efficiency within this range. The diesel consumption is expressed in terms of *Specific Fuel Consumption*—designated as SFC. SFC is weight of diesel consumed in *gms for 1 BHP per hour of running*. Engines with 650-1800 RPM have an SFC of 169 gms/bhp/ hr equivalent to 0.20 lts/bhp/hr, with specific gravity of diesel being 0.835; but on an average, they consume 188 gms/bhp/hr.

Engines with 1800-3600 RPM consume 220 gms/bhp/hr equivalent to 0.26 lts/bhp/hr.

Diesel pumps are selected based on the yield required and the total Head acting on the pump.

Irrigation water requirement:

Every crop needs a certain calculated quantum of irrigation. The total water requirement is computed by summing the peak requisite for various crops grown in any particular season, expressed in liters per second per hectare of crop. This requirement of a crop, known as the Δ delta of a crop is either obtained from specifications similar to table 27, or computed from equation 13. Table 27 shows the water requirement for some of the major crops. The data is just broadly indicative, and not comprehensive. It can be made use of in the absence of detailed agricultural literature on the irrigation cycle of these crops.

This agricultural data, outside the purview of this Book is included more to emphasize the limited quantity of water required for some major crops, to stress need for economy in use of water. The normal practice of flooding agricultural crops results in vast wastage and depletion of water and power. Agriculturists labor under the myth that more water increases production; unless this allegory is dispelled, vast quantities of irrigation water will continue to go waste.

Table 27: Irrigation water requirement for major crops

Irrigation Tables					
Crop	**Crop period in days**	**Number of irrigations**	**Irrigation interval in days**	**Irrigation depth (cms)**	**Water required for 6 hrs. pumping in lps/ hectare**
Paddy	120	17	7	7.5	5
Wheat	120	5	20	7.5	2
Maize	100	7	20	6	2
Bajra	110	4	30	6	1
Groundnut	120	5	20	7.5	1.5
Cotton	120	6	20	7.5	2
Sugarcane	365	6	20	7.5	3.5
Grams	120	3	30	6	1.2
Oil seeds	120	3	30	7.5	1
Mustard	90	5	15	5	1.3

$$\Delta Q = 28\frac{A \times I}{R \times t}$$

Where:

ΔQ = Required discharge in LPS
A = Cropped area in hectares
I = Depth of irrigation in cms
R = Rotation period in days
t = Working hours per day

Example: Wheat needs a depth of irrigation of 7.5 cms, irrigated once every 15 days. If the working hours are 6 per day, required quantum of water is 3.5 LPS assuming that 1.5 hectares of land is under cultivation, calculated as below:

$$Q = 28 \frac{1.5 \times 7.5}{15 \times 6} = 3.5 \text{ LPS}$$

Water requirement computations should take into account the maximum estimated requirement during the successive two years. In dry seasons, water levels may decline to below normal pumping levels. In case of shallow well centrifugal pumps, it is possible to lower the pump assembly into the well to maintain the required suction head (4.5 meters). In case of deep well pumps, the TDH valuation should include dry season depletion levels.

Table28 below is an abstract of the selection chart for diesel prime movers, which lists the recommended HP for various discharges against a range of 'Heads'. The range includes suction, delivery, and friction Heads:

Table 28: BHP requirement of diesel prime movers

BHP	GPM>	36	74	110	150	190	225	260	300	340	375
	LPS>	3	6	8	11	14	17	20	23	26	28
2	Head In Meters	6-18	6-12								
3			13-17	8-11	6-8	6					
5			18	12-18	9-15	7-12	6-10	6-9	6-7	6-7	6
6					16-18	13-15	11-12	10	8-9	8	7
7						16-17	13-14	11-12	10	9	8
8						18	15-16	13-14	11-12	10-12	9
9							17-18	15-17	13-15	13	10
10								18	16-17	14-15	13
12									18	16-18	15
14											18

Computing fuel consumption:

Annual consumption of diesel can be computed from the following equation if a log is maintained of the daily running hours of the pump:

$$\text{Consumption} = \frac{\text{SFC} \times \text{BHP} \times (\text{hours / year})}{0.835 \times 1000}$$

The H.P. of the prime mover is selected on the BHP of the motor required to rise water under known hydrological conditions viz., Q and TDH. Following equation to compute the correct HP incorporates a safety factor of additional 20% over and above the required theoretical BHP:

$$\text{HP of engine} = \frac{12 \times Q \times \text{TDH}}{75 \times e}$$

Where,

Q = discharge in liters per second.
TDH = Total dynamic head in meters.
e = efficiency in %

Points to remember while buying a diesel engine:

1) Performance curves are supplied with all standard makes of diesel engines. The performance curve should show graphically SFC and BHP against RPM rating. The selected engine should have least SFC against required HP and RPM. (RPM is generally 1500 for most shallow well Units).
2) The pump characteristic curve showing its efficiency against RPM should synchronize with the RPM of the prime mover.
3) The BHP of the diesel engine should be equal to the BHP plus 20% additional power.
4) SFC of a diesel engine in the 650 to 1800 RPM range should not exceed 199 gms/bhp/hr, and that of an engine in the 1800 to 3600 RPM range should not exceed 230 gms/bhp/hr.

The lubricating oil consumption of an efficient engine should not exceed 1.5% by volume of diesel consumed.

Selection of electric motors for pumps:

1) Electric motors for pumps are selected as in the case of diesel engines, with the added precaution, that its HP is correlated with the HP of the pump so that it operates as proximate to its full load, or within 15% of its rated full load. An electric motor of HP lesser than that required by the pump, if coupled and operated with a pump requiring a higher HP, would soon overload and burn.
2) Substandard electric motors may commit less initial outlay, but would consume recurring excess power, and will invariably possess a short life span.

Table 29: Selection chart for power driven pumps for agricultural application

HP	GPM>	36	74	110	150	190	225	260	300	340	375
	LPS>	3	6	8	11	14	17	20	23	26	28
0.75	Total Head In Meters (average)	6									
1.00		7/9									
1.25		10									
1.50		12	6								
2.00		16	8	6							
3.00			14	8	7						
4.00			17	12	9	7	6				
5.00				16	14	12	9	7	6	6	6
7.5					17	15	13	12	10	9	8
10							17	16	14	12	11
12.5									18	17	16
15											18

Selection of pumps for domestic application:

Selection of a proper water system for a specific application depends on two basic factors:

1) The water requirement.
2) Quantum of water available.

For domestic consumption, including kitchen, bath and toilet 200 litres per day per person is considered the daily average.

An 11.5 cm (4.5") diameter well normally provides sufficient water to satisfy the requirements of an average household. A power or diesel operated jet pump provides the optimum pumping device. Automatic systems switch on and switch off the pump at pre-set storage levels, ensuring a sustained supply. For obtaining optimum output from the well, pump selection should be based on the known yield of the well measured during drilling (plus 40%), and the Head, which is the vertical distance from the lowest pumping water level to the point of discharge plus friction along pipe lines.

Selection of pumps for industrial application:

Areas selected for locating industries, which may be devoid of surface water resources, and which are to depend to a large extent on groundwater, need extensive investigations. Investigations include detailed geohydrological mapping, exploratory drilling, aquifer performance tests to determine transmissibility, specific yield or storage coefficient and analysis of annual rainfall data to establish groundwater balance between withdrawal and replenishment.

Best means of obtaining large quantities of groundwater to sustain industrial production is through intelligently located, 15 to 30 cms diameter tube wells, and installing suitable turbine or submersible pumps. The pump selection is limited by the well diameter. There are submersible pumps to suit 10, 15, 20, 25, and 30 cm diameter wells. The pump impeller designs range from radial flow with high pumping Heads and low output, to mixed and axial flows with high discharge under lower Heads. The number of stages determines the TDH of the pump. Generally, the number of stages multiplied by the rated Head of one stage indicates the total Head capacity of the Unit.

Methods to maintain system pressure:

The gravity tank method is the most economical means of making water available under pressure in a water supply system, and for providing a stored reserve. Direct pumpage is another means of maintaining pressure in the system. Pumps are connected either in series, or in parallel to maintain a constant supply and pressure in the pipes. Pumps operating in such a system are usually electronically controlled, to switch on and off, in a regulated sequence. The timing and sequence are computed either based on preset periods of operation of each of the pumps or when the pumping level in the wells decline to a specific pre-determined level. Such an arrangement is crucial wherever a large number of wells are working in unison to

1) avert over-pumping,
2) prevent interference between wells, and
3) to sustain a constant flow and pressure in the System.

Parallel operations:

In municipal and some industrial water supply systems involving water circulation, such as in fertilizer plants, and thermal power generating stations, there occurs a wide capacity variation depending on demand. Parallel operation of pumps is required to overcome these fluctuating demands on water. Two or more pumps are hooked up, operating in parallel through a common manifold. Two pumps deliver water continuously, while the third pump operates intermittently to supplement additional demand.

Following features should be ensured while selecting pumps for parallel operations:

i) Pumps, pumping in parallel should be of similar characteristics;
ii) Head-Capacity curves should be either rising, or flat. In case pumps in parallel operation display a drooping characteristic, then it should be ensured that while starting the pumping operations, no pump is running at its shut-off Head.
iii) Pumps with dissimilar characteristics can run in parallel, if the system Head does not exceed the shut-off Head of any pump,

at any capacity, resulting from the combination of other pumps operating in the System.

Pump auxiliaries:

Following accessories are installed along with the pump to ensure optimum service life for the unit:

1) For all deep well installations, a non-return valve should be coupled to the top of the pump.The manufacturer will indicate the type of non-return valve best suited to the particular pump.
2) A dry-run-preventer switches off power supply, stopping the pump, whenever the water level in the well is depleting to levels, which may cause the pump to break suction. A pump tending to break suction portends approach of lower than minimum required water levels in the well. Submersible pumps running dry, (with insufficient NRIH and without minimum water column above the entry duct), will sooner or later result in seized bearings, and burn.
3) Suction shrouds should be installed to ensure proper cooling of the motor.
4) An automatic Starter, with a thermal overload relay, and a built-in ammeter, to protect the installation against power fluctuations. Variations in amperes drawn by the motor are an early warning of approaching problems with either the pump or the motor.
5) A Voltmeter is necessary to monitor the voltage. The Unit should not be started, if the voltage is below 390, or above 450.
6) A Single-phase preventer, calibrated to the H.P. of the pump, should form an integral part of the control panel. In the event of any one phase cutting out, the motor will draw its energy requirement from the remaining two phases, with an enhanced load of 1.73 times the normal, causing the motor to burn. Single-phase preventers with a built-in time delay, would circumvent tripping under transient non-feed conditions. They should also have been designed to protect the motor against unbalanced conditions, which may arise both upstream and downstream of the motor.

Wells sourcing groundwater for industries:

Industries that depend on groundwater for their production facilities need to exercise most important care in survey and exploration for location of accurate well points. Wells drilled at these points should have the potential to yield and sustain supply of large quantities of water without interruption. Any disruption in yield, because of inappropriate location of the well will have cascading lethal effect on the manufacturing process and the final product. Investigations for industrial wells will necessarily comprise details of the recharge area, infiltration, run-off, transmissivity, safe yield and similar such aspects to equate the balance between groundwater extraction and replenishment.

If well points have been carefully selected and tested, large quantities of groundwater can be drawn by means of 15 to 30 cm diameter deep wells by installing appropriate submersible or turbine pumps. Aquifer performance tests are a significant pre-requisite for selection of pumps for heavy-duty wells.

Costing pumping operations:

It is desirable to monitor regularly the cost of water that is being pumped from any well or an array of wells. Such costing is essential in town water supply systems and industries deploying groundwater for product manufacture. If a pumping system that is inappropriate to the specific hydraulics of the water source has been installed, it would show up in the costing. Ignoring costing may accrue unfructous cost of the product. Expenditure in excess of the outlay often signify that the pumps installed are not calibrated to the hydraulic characteristics of the source. Even modifying the pump placement in the well would often result in considerable reduction in power consumption, increased pump efficiency, and significant increase in the quantum of water that is being obtained from the well. Cost of pumping can be computed from the following equation:

These equations compute the approximate operating costs per hour of operation, without taking into account the initial capital, or the depreciation cost of the plant, maintenance and man-hour costs of the operating the system.

Electric motors:

$$\text{Cost per hour of operation in Rs.} = \frac{Q \times H \times 0.746 \times kw}{4569 \times e \times em}$$

Where:

Q = Discharge in liters per minute (LPM)
H = Total dynamic head in meters
kw = Cost of kw/hr in rupees
e = Overall pump efficiency in %
em = Motor efficiency

For example, where,

Q = 150 LPM (2000 GPH)
H = 25 m
kw = Rs. 2.75
e = 75%
em = 90%

$$\text{Cost per hour of operation in Rs.} = \frac{150 \times 25 \times 0.746 \times 2.75}{4569 \times 0.7 \times 0.9} = \text{Rs. } 2.70 \text{ / hour}$$

Diesel motors:

$$\text{Cost per hour of operation in Rs.} = \frac{Q \times H \times F \times R}{4569 \times e}$$

Where:

Q = Discharge in LPM
H = Total dynamic Head in meters
F = Fuel consumption per HP per hr in litres
R = Price of diesel per liter in Rs.
e = Pump efficiency

For example, where:

Q = 150 LPM
H = 25 m
F = 0.25 liters / hour
R = Rs.50.00
e = 70 %

$$\text{Cost per hour of operation in Rs.} = \frac{150 \times 25 \times 0.25 \times 50}{4569 \times 0.7} = \text{Rs. } 14.70 \text{ / hour}$$

(modified from the original calculations by Layne & Bowler Inc., USA)

CHAPTER XIII

INSTALLATION OF PUMPS AND ANCILLARIES

Surface centrifugal pumps:

Proper positioning and installation of centrifugal pumps is an essential factor of any water-pumping project. Improper installation results in drastic drop in pump efficiency, output less than the rated capacity, high-energy consumption, and undue wear and tear on the pump, and its ancillaries.

Following aspects, during installation of a surface centrifugal pump, are of particular importance, and need exacting attention:

1) Location; 2) Foundation; 3) Alignment; 4) Discharge assembly and 5) Suction assembly.

1) Location: The pump should be positioned as close to the well, or the supply tank as feasible to ease suction lift to the minimum. Suction lift should be limited to not more than 5 meters; in case the water is at a marginally greater depth, pump should be placed at lower levels, either on a ledge built on the well wall, a cavity formed in the wall or a pit dug close to the well.
2) Foundation: Permanent, or semi-permanent installations should be bolted to a base plate, which in turn is secured to a concrete foundation. System should be free of any vibrations. While fastening the pump to the foundations, it should be ensured that load and stress on the base plate is even and uniform. Unbalanced strains will result in bearings running out of alignment, causing undue wear, and inefficient running of the pump.

3) Alignment: Belt driven Units should be so located that the pulley of the driver, and the pulley on the pump are in exact alignment. The pump is placed on wedges, by manipulation of which, alignment is ensured. The shaft should rotate freely when turned by hand.
4) Delivery assembly: The suction and discharge pipes should be compatible with the pump suction and discharge terminals, and with the pumping capacity of the pump. Table 30 is an abstract table briefly tabulating the specified suction and discharge pipe diameters for diverse discharges:

   The delivery pipe should be independently supported near the pump, without any part of its weight being shared by the pump casing. It should be as short and direct as possible to reduce friction Head. Gate valves should be installed at the pump suction and discharge; these valves offer no resistance to flow, while providing a tight shut-off. It is advisable to connect the pump with the discharge pipe through a concentric increaser to minimize loss of Head. A check valve installed close to the pump protects the pump from any backpressure, and prevents reverse flow in case of power failure. Gate valve shuts off flow. It is kept shut during pump priming, and when the pump is shut down. The power required to drive the pump is proportional to the flow; hence, flow in excess of requirement means unnecessary use of excess power.
5) Suction assembly: The suction assembly should suit the volume of flow being pumped, and should be about one and half times larger than the pump inlet, connected to the pump by means of an eccentric reducer. The pump should be located as close to the water source as feasible; total dynamic suction lift should be limited to less than 5 m. The suction pipe should be installed with an eccentric reducer (fig.31) whenever a pipe size transition is required. The flat of the reducer should be on top when the water is being drawn from lower or same level; and at the bottom, if the water is coming from a higher level.(fig.29). This will avoid build-up of an air pocket at the pump suction and allow air to be evacuated automatically

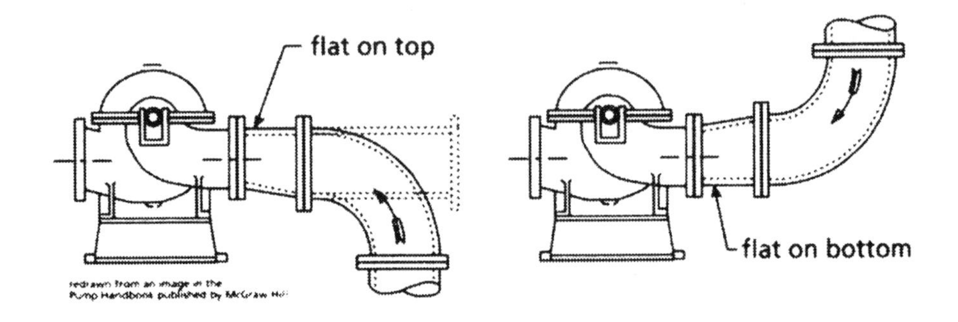

Fig 29: Deployment of 'flats' near the pump avoids turbulence.

Table 30: Recommended diameters of suction and delivery pipelines

Discharge		Suction		Delivery	
GPM	**LPS**	**mm**	**inches**	**mm**	**inches**
36	2.8	50	2	40	1.5
74	5.6	65	2.5	50	2
110	8.4	80	3	65	2.5
150	11.2	100	4	80	3
190	14.2	125	5	100	4
225	16.9	125	5	100	4
260	19.8	125	5	100	4
300	22.7	150	6	125	5
340	25.5	150	6	125	5
375	28.3	150	6	125	5

The suction assembly should include least number of Bends and Elbows, to decrease the friction Head; and wherever installed should encompass a long radius. If it becomes necessary to incorporate long horizontal suction pipes, they should slope-up at an angle of about 20° from the supply to the point of connection with the pump. The suction assembly should be taken below any obstruction, instead of above the obstruction to prevent air pockets that tend to collect at all high points in the system. An air pocket in the suction will invariably break the water flow. It is for this reason that a concentric reducer should not be included in a horizontal suction line. The reducer should be eccentric with its flat side at the top (fig.31) so that flat tops of the larger and smaller pipes are both at same level. A standard straight taper adopter would allow air pockets to form at the top of the larger pipe. The entire suction line should be self-supporting, without any strain on the pump to which it is connected.

Following are some of the important points to be observed while connecting the suction and delivery pipe lines to surface centrifugal pumps, for pumping water from open wells and tanks:

1. In the suction line, elbows and suction pipes should be either vertical or slope gradually downward; (fig.31).
2. In the discharge line concentric increaser should connect with the check valve followed by the gate valve; (fig.30).
3. Suction terminal assembly should have sufficient submergence well below the pumping water level of the well, minimum being 1 meter.
4. Wrong suction and discharge assemblies will result in accumulation of air bubbles which will reduce output leading to cessation of pumping.

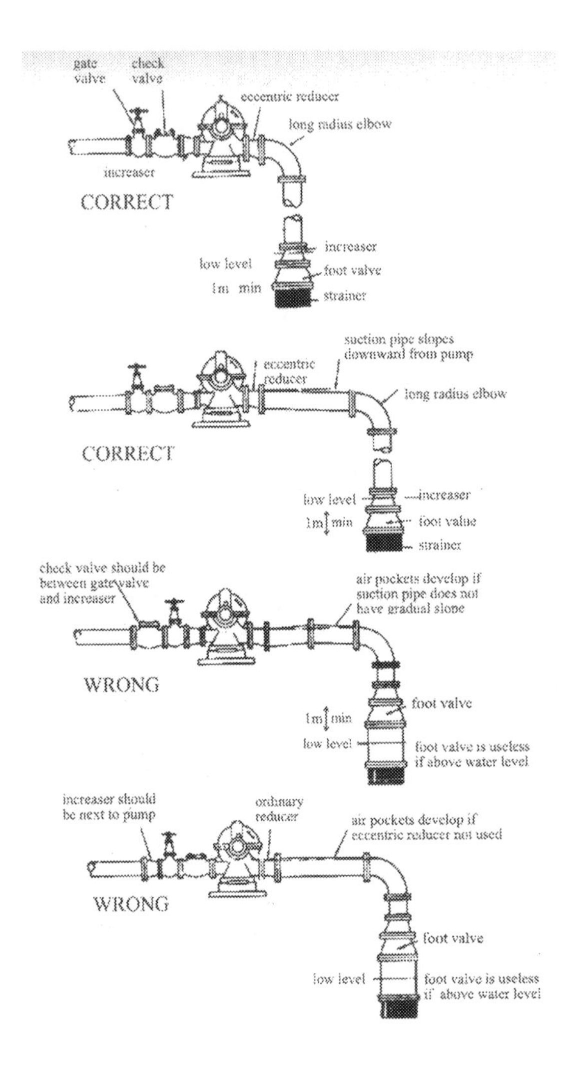

Fig 30: Correct and incorrect ways of installing suction and delivery assemblies
(Original depiction by Cameron Pump Division—Ingersoll Rand Company. Courtesy: Pump Handbook by Volney C.Finch

Figure 30 is a graphic illustration of the "correct" and "incorrect" ways of

1) Connecting suction and discharge pipes,
2) Positioning of regulating valves; and
3) appropriate positions of eccentric and concentric adopters.

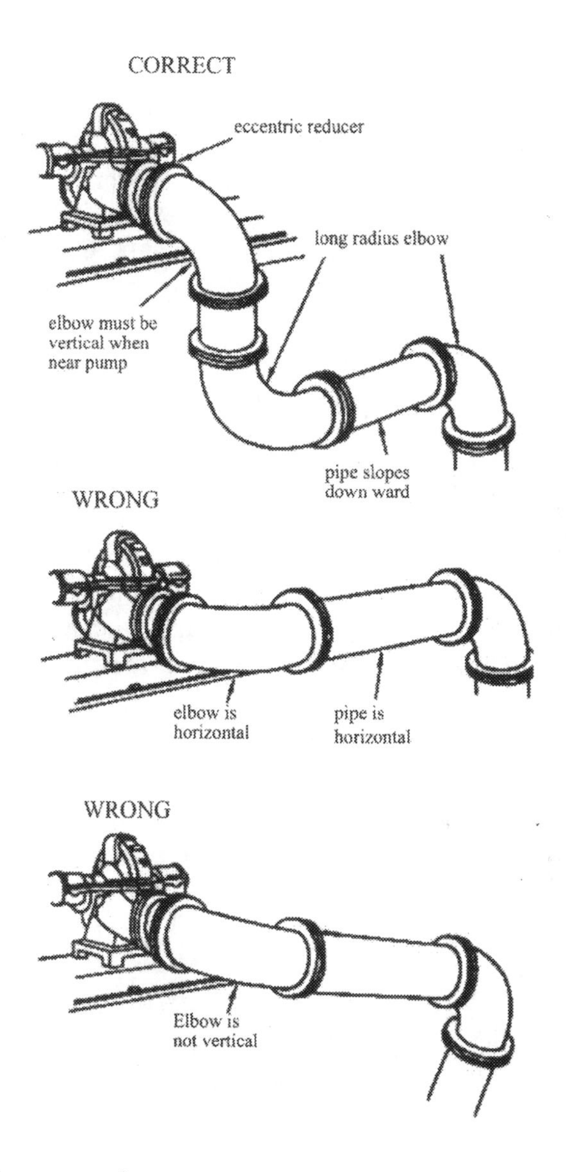

Fig 31: Correct and incorrect ways of installing suction piping. (Original depiction by Cameron Pump Division—Ingersoll Rand Company. Courtesy: Pump Handbook by Volney C.Finch

Figure 31 illustrates graphically the correct and wrong positioning of connecting ancillaries in the suction line.

Jet pumps:

Jet pumps consists of two major Units: the jet assembly inserted into the well, below the lowest water level, and the centrifugal pumping unit on the surface which actually draws water from the well. The installation consists of two pipes, one to force water through a nozzle and the venturi into the well, and the other pipe to pump water out of the well.

The pumping Unit on the surface should be installed in a clean dry location, with proper ventilation to avoid over heating of the Unit. The pumping Unit can be installed directly on the well Head, close to the well, or any convenient distance (not exceeding 60 m)from the well. This length will, however add to the Head acting on the pump resulting from friction in the additional length of the pipelines. The pipes leading from the pump to the well Head must be horizontal, or slope down to the well with a gradient of about 1 in 50. New pipes and fittings should be deployed to avoid scaling, and dirt from old pipes, which may clog the sensitive nozzles and venturi. These additional lengths are clamped to pipes inserted into the well, which could be of either HDPE or GI. Installation of GI pipes require a hoist and tripod, while HDPE pipes could be manually installed. HDPE or similar plastic pipes cause less friction for the flow; minimize Head on the pump, resulting in lower power consumption.

Submersible pumps:

Most of the submersible pumps are designed to fit 114, 152, and 165 mm wells. However, there are pumps that are manufactured for deployment with 250 mm diameter wells. The Horsepower determines the power of a pump, but not its ability to deliver any specific quantum of water. This quantum, in addition to HP, is determined by other factors. It is a myth that high HP pumps pump more water. The pumps are designed either for high yield, or high Head, but not for both. Head and yield are inversely proportional to each other. The Head capacities of the pumps depend on the number of stages incorporated into the system. The ultimate power and ability of the multi-stage pumps is further determined by the design width of the impeller and diffuser, and not by the number of stages alone. The pressure obtained from the pump is a function of the diameter of the impeller, number of impellers, and the speed at which it rotates. The Head (Read pressure) developed is proportional to the square of the speed, and square of the diameter of the impeller. They operate nominally at 3450 RPM, the fastest feasible speed for a 60-hertz electric motor.

There are both single-phase 230 volts, and 3-phase 440 volts submersible pumps. Single-phase motors are inherently not self-starting. The auxiliary starting devise consists of a starting capacitor connected to a start winding of the motor. The capacitor gets the motor started, and runs it up to running speed following which a switch disconnects the capacitor and the start winding, and the motor runs on its own windings. The disconnecting switch is generally a relay placed on the ground in a control box together with the starting capacitor, an overload protector, and a low water level cut-off. Installation of a capacitor is important, as it smoothens the running of the pump and reduces power consumption. Control box including the 'earthing' terminal has a three-wire connection to the motor, which should not be mistaken for a 3-phase operation. The systems in the control box match the voltage, HP, and starting characteristics of the motor with which it is supplied, and are not interchangeable.

Three-phase motors are self-starting with conventional starters. Control boxes contain the starter switch, a single-phasing preventer, overload protector, and a low water level cut-off. These starters being sensitive to heat should be protected from direct sun light, and extremes of heat and cold. A voltmeter indicating the voltage of the power feed, and single-phasing preventer to reduce motor burn-out, and a low water level cut-off are essential ancillaries on the switch board.

Submersible pumps are concomitantly a fragile, resilient, and sturdy machine that works completely submerged, in a hostile environment. It is set in a narrow working space, with no tolerance limits, under constantly varying pressure conditions and fluctuating power load. A superior pump with a capacitor and reliable service back up are essential basics while purchasing a submersible machine. Such a pump, with a capacitor, and a HDPE or RPVC eductor with a friction-less foot valve ensure an economic and long lasting water pumping system.

Before installing the pump, it is good practice to chlorinate the well by introducing HTH (Chlorine) into the system.

Installation procedure:

1) A tripod with pipe wrenches and clamps are essential tools while installing GI pipe-lines, while HDPE pipes can be manually installed without heavy equipment. Lowering a 12 mm air-line parallel to and clamped to the eductor pipe helps in checking water levels (both static and pumping water levels) whenever required. Periodic checking of water levels is a prognostic tool in verifying reasons for any short fall in water out-put. A deep well may cease to yield water for many reasons, other than the well drying-up.
2) Before lowering, ensure that the pump is properly and firmly screwed to the eductor pipe.

   Pump should be lowered to the pre-determined level, taking into account the maximum draw-down and the NPSH and NRIH values of the pump.
3) After reaching the placement level, adequately stuffed double clamps placed at right angles to each other, or a tee, should be firmly secured to the top of the eductor, and placed so as to rest smoothly on the wellhead.
4) After ensuring firmness of the clamps and the drop-pipe, seal the well, and run the power cable through the vent in the seal. Connect the cables to the power mains, via the control box with appropriate earthing and insulation.
5) Ensure that the positive and negative terminals of the pump are appropriately connected to the respective 'load' terminals of the magnetic starter. If the connections are incorrect, the motor will rotate in the opposite direction yielding little or no water. In such a case, instead of suspecting the pump, or the well for the low yield, interchange any two of the cable leads (positive to negative, and negative to positive) on the 'load' terminals of the magnetic starter. It is dangerous to run a 3-phase motor backwards for more than one minute.
6) Install a throttle valve with a pressure gauge on the discharge arm of the eductor. This would help balance discharge and recharge, by partially shutting off a 'weak' well. In case of such a weak well pumping to more than its potential, as the discharge drops the gauge will indicate a declining pressure. This reducing output should be balanced to yield a more steady

discharge by partially closing or opening the throttle valve. To avoid tampering with the balanced setting of the throttle valve, fix a valve with a square head stem with a lock nut.

7) As the pump runs, test the well water for clearness. Run the pump until the water runs clear. Shutting off the pump while the discharge water is not clear of any sand or silt might cause the system to sand-lock leading to choked connections. Pump should be run until the water is clear.

As the well continues to pump water, the yield would gradually increase, sometimes over a course of several months, until the system attains its optimum potential.

Pump priming:

Any pumping system entails creation of vacuum in its suction line, casing, and impellers before it is able to draw water from its source, and convey it to the point of discharge. This process of creating a vacuum in the line is called 'priming'. The submersible, the turbine, and the jet pumps are self-priming, while the shallow installations need priming before starting. One of the simplest ways of priming a system is to provide a foot valve at the suction inlet; shut the discharge valve; and allow water from a secondary source to run into the suction pipe and casing, through the air valve till it overflows from the pump casing. In systems where an automatic priming valve is provided, the priming water is filled into the pipelines until the automatic valve closes. Frequently priming water can be drawn from the discharge itself, wherever the discharge lines are short enough without a check valve. In such a system, it is necessary to close the discharge valve before shutting off the pump; and before starting the pump, discharge valve is opened to allow the water to run back through the pump into the suction assembly. Large centrifugal pumping systems need ancillary vacuum pumps, which provide the most satisfactory priming. Foot valves are not required when vacuum pumps are provided for priming; and when a good vacuum pump, with a vacuum regulator is employed, there is very little risk of starting an unprimed system. As an additional measure of safety, the vacuum and the main pumps are often interlocked.

Starting any centrifugal system, with insufficient water in the assembly is a hazard for the pump parts, which are water lubricated. Running dry even for a few seconds will result in undue wear, and running dry for a longer period will generate sufficient heat to make the wearing rings seize.

CHAPTER XIV

PIPING SYSTEMS

Water from the source to the point of usage is through pipelines, whose size and distinctiveness is decided based on the volume of water being conveyed. There are very many types of pipes, ranging from steel to plastic. Most common is the GI pipes, but they are rapidly going out of use. The galvanizing coating, commonly zinc, prevents the rusting of the iron pipe. However, overtime and with use, the galvanizing solution will corrode, causing the metal beneath to rust. Minerals in the water supply contribute to the corrosion, causing the pipes to clog with corroded debris. Inner rusting will not be visible until water starts seeping out.

Pipes, whether of galvanized steel or PVC, have limited capacity to carry water economically, without causing turbulence in flow.

Polybutylene (PB) and cross-linked polyethylene pipes are two similar water pipes that come in rolls. They have the advantage of lightweight, flexibility and are durable, inexpensive, and easy to install. They require mechanical fittings and cannot be soldered. However, they can be manipulated as required during installation, obviating necessity of elbows, joints and other couplings minimizing labor costs.

Copper pipes are a frequent alternative, as they present an elegant exterior. They are expensive, are totally weather proof and long lasting, but require skilled labor for installation.

Chlorinated polyvinyl chloride (CPVC) pipes offer a cheaper alternative. They can be joined together by crimp ring fittings, PVC fittings or compressor fittings.

High-density polyethylene (HDPE) piping is by far the most versatile and cost-effective means of transiting water either over-ground, buried under-ground, floating or in sub-marine situations. It has the following advantages:

1) Is strong, extremely tough, does not rust or corrode and is maintenance free.
2) Light in weight and easy to install.
3) Heat-fused joints create a homogeneous, monolithic system; the fusion joints being often stronger than the pipe itself.
4) High strain tolerance limits withstand surge and water hammer action.
5) Maintains smooth flow rates, does not tuberculate.
6) Installation costs are minimal due to its longer length and flexibility
7) Allows bending radius of 20 times the outside diameter of the pipe.

HDPE pipes provide a highly cost-effective method of pumping water from tube wells as, in addition to being flexible and strong, are exfiltration and infiltration proof, precluding contamination from subsurface and vadose zones. For deployment in tube wells, the HDPE pipes have the following advantages and disadvantages.

Advantages:

1) Cost effective, and being a major component of the pumping system significantly trims the overall cost outlay on any well.
2) are flexible and maneuverable during lowering of the line into the well, reducing the probability of being obstructed by unevenness of the well walls, or in wells which may have been drilled off the vertical.
3) being light in weight, can be lowered manually without the need for heavy lowering equipment.
4) can be easily pulled out for inspection, cleaning footwalls or clearing the well of any sand filling.

Disadvantages:

1) These pipes being made of high density polyethylene are flexible and elastic and are capable of withstanding rough environments and handling; but carelessness and negligent rough handling could result in a tear or puncture. An incipient tear caused during lowering or joining the pipe ends will result

in air leak and a break in the vacuum leading to cessation of water flow.

2) HDPE pipes are not fit for use with high Head 150 mm submersible pumps. If used, should be checked at the first sign of lessening output of water. Since the pipe bears the weight of the submersible pump, the pipe manufacturer's recommendations are essential for the suitability of the pipe for use with heavy duty, multi-stage pumps.
3) They can be fitted to 4" submersible pumps, if they possess 10-12. kg/cm^2 pressure rating and can be deployed to a maximum depth of 260 m (850 ft).
4) HDPE pipes with a pressure rating of minimum 4 kgs/cm^2 can be installed with jet pumps down to moderate depths; while for use with ½ to 2 HP multistage jet pumps, the pipes should have a certified pressure rating of 6 to 8 kgs/cm^2 when they can be installed to a maximum depth of 150m (500 ft). HDPE pipes are available in 20 to 1600 mm diameters with calibrated flanges capable of withstanding pressures of 2.5 to 16 kg/cm^2 (35.5 to 227.5 psi) internal pressures.

Objective of any water transmission system is to move the water smoothly, with as little turbulence and less friction as possible. Routing the pipeline is significant. Air accumulates at high points. An ideal pipe run is one, where the piping gradually slopes up from the pump to the outlet. This will ensure that any air in the discharge side of the pump can be evacuated to the outlet.

Water hammer:

Water hammer is a pressure surge or wave resulting when water in rapid motion is forced to stop or change direction suddenly. Water hammer commonly occurs when a valve is closed suddenly at an end of a pipeline system, and a pressure wave propagates backwards in the pipe. This pressure wave can cause major problems, from noise and vibration to pitting and pipe collapse. It is possible to reduce the effects of the water hammer pulses with accumulators and other features.

Water hammering is particularly serious in large piping systems such as are installed in municipal water supply distribution systems. Long, upward inclines followed by downward sloping pipes characterize these systems. Solutions to this can involve special pressure/vacuum reducing valves at the high and low points or additional tanks, which provide a buffer for pressure surges. (See http://www.ventomat.com/default.asp).

Water flow through a pipe or a system of pipes is governed by certain physical limitations, which determine the volume and velocity of flow. Velocity of flow is determined by:

$$V = 0.4085\left(\frac{Q}{2d}\right)$$

Where:

V = Velocity in feet/second
Q = Volume in gallons/minute
d = Inside diameter of pipe in inches

OR

$$V = 1.274\left(\frac{Q}{2d}\right)$$

Where:

V = Velocity in meters/second
Q = Volume in meters^3/second
d = Inside diameter of pipe in meters

Rate of water flow in pipes, and conduits, and the resulting friction Head losses are major factors to be taken into account, while computing total dynamic Head acting on the pump. When water or any fluid transits through a pipe, it has to overcome resistance caused by friction, ensuing from both its movement through the pipe, and its own turbulence caused by its flow under pressure. For this reason, if pressures were recorded in a length of pipe, the pressure gauge at the outlet would register a lesser pressure than the one at the inlet.

As the rate of flow increases, there occurs a proportionate drop in pressure. If the rate of flow doubles, the pressure drop increases four times. Small diameter pipes record higher friction loss while large diameter pipes develop lesser friction loss for the same quantum of flow. Doubling the diameter of a pipe enhances its capacity of the pipe by four times. (Not two times.) Table.33 is an abstract of the diameters required for suction and delivery pipelines for diverse discharges. For any pumping system, diameter of lines required for suction are larger than those required for delivery.

Computation of friction 'Head':

Power is required either to lift water to a higher level, or to push it against friction through pipes. The quantum of power required depends on the rate of flow and the pressure or Head against which the water is being pumped. Since additional power required to overcome friction is not available, this actual loss in velocity accrues on the pump, and is called friction Head. This additional Head, caused by friction is taken into account, and added to the over-all Head, while selecting the pump for the well. This friction Head or loss in quantum of flow occurs in both suction and delivery pipes, enhanced further by the number of bends, elbows, and valves in the conduit system. This Head loss varies with the flow rate, pipe diameter, type of pipe, and the interior condition of the pipe.

Hazen & Williams equation:

The Hazen & Williams is an empirical equation that relates the flow of water in pipes, with the physical properties of the pipe, and the pressure drop caused by friction between the flow and the pipe walls. The equation lacks theoretical validation, but is widely used to estimate the velocity of flow, and to compute Head losses that occur in pipelines conveying water at normal temperature and viscosity. It is not valid for other fluids. Velocity is computed from the equation:

$$V = 0.849C \times R^{0.63} \times S^{0.54}$$

Where:

V = Velocity in meters per second
C = A constant which varies with the smoothness of the pipe (Hazen-William constant)
R = Hydraulic radius in meters;
S = Hydraulic gradient (head loss) in meters per meter.

The hydraulic radius 'R' is the cross sectional area of flow, divided by the wetted perimeter of the pipe. Thus for a pipe flowing full R = D/4, where 'D' is the pipe diameter. Since the rate of flow equals the velocity times the area of flow:

$$Q = VA = 0.849C \times A \times (D/4)^{0.63} \times S^{0.54}$$

Where:

Q = Flow in cu meters per second
D = Diameter in meters
S = Hydraulic gradient in meters per meter:
A = Cross sectional area of the pipe ($\pi d^2/4$)

To determine 'Q' in liters per second or liters per minute

$Q(L/Sec) = 278.5C \times 2.63D \times S^{0.54}$
$Q(L/min) = 16710C \times 2.63D \times S^{0.54}$

Table 31: Values of Hazen-Williams Constant 'C'

Conduit material	Recommended values	
	New pipes	Design value
Cast iron	130	100
Galvanized iron 50 mm	120	100
Steel, with riveted joints	110	95
Steel, welded joints lined with cement	140	100

Conduit material	Recommended values	
	New pipes	Design value
Steel, welded joints	140	110
Concrete	140	110
Asbestos cement	150	120
Copper	140	120
Polyethylene	140	140
PVC	130	130
Fiber-reinforced plastic	150	150

The Darcy-Weisbach equation to estimate pressure loss in pipes resulting from friction, has as its basis a more accurate model, but has an iterative approach, and hence not preferred.

Pressure loss due to friction in pipelines is more easily obtained by the Hazen-Williams equation:

$$f = 10.67 \times L \times Q^{1.85} / C^{1.85} \times dn^{4.87}$$

where:

f = pressure loss over a length of pipe, in meters
L = length of pipe, m (meters)
Q = volumetric flow rate, m^3/s (cubic meters per second)
dn = inside pipe diameter, in meters.

For most purposes, the Hazen & Williams equation is best stated and applied from a pre-calculated format, which gives pressure loss accruing from friction in straight and level pipes of various diameters in respect of diverse volumes of flow.

The values of head loss abstracted in the Tables 32, 33 and 34 are broadly applicable to pipes in which the roughness protrusion on the pipe walls are completely submerged in the laminar sub-layer, which is the very thin coating at the pipe wall in which only laminar friction exists.

Table 32 abstracts the Head loss in meters for small to medium flows (300 to 9000 GPH) in plastic pipes for a length of 100 m. Friction in plastic pipes made of HDPE, PP, PE and similar material is less than in CI and MS pipes, and should invariably be preferred, for all long-term and major Projects.

Table 33 is to compute Head loss for every 10 meters of metallic pipes.

Flow in pipes other than given in these tables, can be either computed from the equations given above, or can be obtained from Engineeringtoolbox.com.

Table 32: Frictional Head loss for every 100 m length of straight plastic pipes

		Nominal pipe diameter									
Inches >>		0.5	0.75	1	1.25	1.5	2	2.5	3	4	6
mm>>		12.7	19	25.4	32	38	50.8	57.15	77	104.5	152.4
LPS	GPH	Friction factor									
0.36	300	20.7	5.3	1.6	0.4	0.2					
0.43	360	29.0	7.4	2.3	0.6	0.3					
0.58	480	49.5	12.6	3.9	1.0	0.5	0.1				
0.72	600	74.7	19.0	5.9	1.6	0.7	0.2	0.1			
1.44	1200		68.6	21.2	5.6	2.6	0.8	0.3	0.1		
2.16	1800				11.8	5.6	1.7	0.7	0.2		
2.88	2400				20.1	9.5	2.8	1.2	0.4	0.1	
3.60	3000					14.4	4.3	1.8	0.6	0.2	
4.32	3600					20.1	6.0	2.5	0.9	0.2	
5.04	4200						7.9	3.3	1.2	0.3	
5.76	4800						10.2	4.3	1.5	0.4	
6.48	5400						12.6	5.3	1.9	0.5	
7.20	6000							6.5	2.3	0.6	0.1
9.00	7500							9.8	3.4	0.9	0.1
10.8	9000								4.8	1.3	0.2

(Courtesy: www.EngineeringToolbox.com—partially modified)

Table 33: Frictional Head loss for every 10 meters length of straight pipe

Flow in LPM	40	50	65	80	100	125	150
	Frictional head loss in meters per 10 meter length of pipe						
2.8	1.91	0.60	0.15	0.05	0.01		
3.0	2.20	0.67	0.18	0.06	0.02		
3.5	3.00	0.91	0.24	0.09	0.03		
4.0	3.88	1.18	0.32	0.11	0.04		
5.0	6.07	1.84	0.50	0.16	0.05	0.02	
5.5		2.20	0.57	0.20	0.06	0.02	
6.0		2.65	0.74	0.24	0.07	0.02	
7.0		3.60	0.92	0.31	0.10	0.03	
8.0		4.66	1.21	0.38	0.13	0.04	
8.5			1.34	0.44	0.14	0.05	
9.0			1.47	0.49	0.16	0.05	
10.0			2.11	0.71	0.22	0.07	
12.0			3.04	1.02	0.31	0.10	0.04
14.0			4.06	1.34	0.43	0.13	0.05
16.0			5.31	1.80	0.55	0.17	0.07
18.0				2.31	0.71	0.22	0.08
20.0				2.44	0.77	0.25	0.09
22.0				2.96	0.94	0.30	0.11
24.0				3.50	1.12	0.35	0.13
26.0				4.04	1.31	0.42	0.16
28.0				4.80	1.50	0.50	0.18
30.0					1.92	0.60	0.23

(original calibration by: Fairbanks & Morse)
Based on IS:2951—I with n = 0.1525

Table 34: System auxiliaries and their resistance to water flow in terms of equivalent length of straight pipe Based on IS-II-1965

Pipe size (mm)	Equivalent pipe lengths in meters				
	Elbow (n=0.7)	90°bend (n=0.19)	Standard T (n=0.4)	Gate valve (n=0.12)	Foot valve (n=3.5)
25	0.536	0.306	0.704	0.077	2.04
40	0.997	0.569	0.131	0.142	3.05
50	1.296	0.741	1.704	0.185	3.96
65	1.814	1.037	2.384	0.259	5.18
80	2.241	1.281	2.946	0.320	6.10
100	2.959	1.691	3.889	0.422	8.23
125	4.037	2.307	5.306	0.576	10.00
150	5.125	2.928	6.735	0.732	12.20

Table 35: Maximum volume of water that can be pumped through 30 m of metallic pipe of various diameters under differing pressures

Pipe diameter (mm) >>	10	20	25	30	40	50	65	75	100
Pressure (kg/ cm^2)	Liters per minute								
1.2	12	35	70	125	195	400	757	1100	2230
2.0	19	52	105	196	295	605	1165	1650	3350
2.8	22	60	124	227	340	696	1324	1905	3870
3.5	25	66	140	265	382	780	1475	2134	4325
4.2	26	74	150	287	416	855	1627	2335	4740
5.2	28	83	170	320	465	957	1816	2610	5300
7.0	34	94	196	374	537	1105	2112	3016	6083

(modified from original calibration by Fairbanks Morse &Co., USA)

The following procedure sequences the steps to compute frictional Head accruing in water conduit systems:

1) Measure the length of pipeline and its internal diameter
2) Determine the number and nature of Bends, Tees, Elbows, and valves
3) The friction accruing from the above pipe fittings is determined from equivalent length of piping from Table 34.
4) Add this value to the total length of piping to obtain the equivalent length for this arrangement of the piping system with its auxiliary fittings. This notional length is 'L'
5) Determine the quantum of water to be pumped in liters per second
6) Refer to Table36 to obtain pressure loss for that flow.
7) This value is designated as 'H', which is the frictional resistance against which the water will flow. Additional Head created by this resistance is obtained by = LH/ 100.

For short length pipelines, in addition to reference to the friction Head Tables, it is necessary to observe the following guidelines:

1. While selecting the delivery and suction pipes, it should be ensured that the size of the suction pipe is such, that the total frictional losses on the suction line do not exceed 0.5 m.
2. The size of the discharge pipe should be such, that the total frictional losses on the delivery line do not exceed 0.82 to a maximum of 1.00 meters for 10-meter length of pipe.
3. The velocity of flow in the suction pipe should be limited to 1.5 m/sec; and to 2.0 m/sec in the delivery pipe.

Table 36: Computing total dynamic head

Pipe diameter in mm>>	80	100	125	150
Velocity in m/sec>>	2.85	1.82	1.18	0.81
Fittings	**Equivalent length of straight pipe**			
Foot valve-1 (A)	6.10	8.23	10.00	12.20
Gate valve-1 (B)	0.32	0.42	0.57	0.73
Elbows—5 (C)	11.20	14.75	21.00	25.00
90 Bend—1 (D)	1.28	1.69	2.30	2.92
Total equivalent length in meters(E1) (A+B+C+D)	18.90	25.09	33.87	40.85
Piping (L=300 m) +equivalent length in meters (L+E1)	318.90	325.09	333.87	340.85
Friction losses in m/10 m length of pipe for 14 LPS (F1)	1.34	0.43	0.13	0.05
Friction losses in equivalent length of straight pipe (hf) (L + E1) × F1/10	42.73	13.97	4.34	1.70
Total static head(hs)	20.00	20.00	20.00	20.00
Total dynamic head in meters (Hd) (hf + hs)	62.73	33.97	24.34	21.70

Selecting a water conduit system:

While selecting a water transmission system, various factors have to be taken into consideration, important of which is the cost factor. Large diameter pipelines reduce friction, resulting in consumption of less power, and lower power costs. This will have a cascading effect, since this economy in power consumption will last through the life period of the project. However, this would require higher capital outlay. Hence, the length of the pipeline and the distance to the water source need careful consideration and a balance drawn between the high initial cost with low power bills, and low capital outlay with high, recurring power bills. The most economical diameter of the pipe in relation to productive flow and cost of pumping are determined by considering a few permutations and combinations of varying pipe sizes, and computing the explicit cost factors. This forms an integral component of any major water supply project, since the objective of sustained flow and the cost factors will be a constant feature throughout the lifetime of the scheme.

Example:A small industrial installation, wholly dependent on groundwater for its operations, requires 14 liters per second (11000 GPH) of water. A single 155 mm diameter well, located in the proximity of an ephemeral surface source (with two additional stand-by wells) drilled to a depth of 50 meters has its static water level at 11 meters. The well has been tested to a dependable yield of 14 LPS for a drawdown of 5 m, pumping under equilibrium conditions. The point of usage, which is at a pipe alignment distance of 300 m from the source, is situated 4 m above the wellhead elevation. Calculations have been made for used pipes with an Allen Hazen 'C' value of 110; and are based on Tables 32 and 34.

Cumulative total static discharge Head is:

Static water level = 11 m
Drawdown = 5 m
Usage point elevation = 4 m
Static discharge Head = 20 m (11 + 5 + 4 m)

The piping system consists of the following features:
Length of pipe line = 300m
Non-return valve = 1
Gate valve = 1
Elbows = 5
90° bend = 1

The ratio of the length of pipeline to the static Head is large. Hence, the diameter of the pipe to conduct 14 LPS over a length of 300 m, and a static discharge Head of 20m is determined by trial and error method. The most economical diameter in relation to productive flow and cost of pumping is computed by considering permutations and combinations of varying pipe sizes.

To install a cost-effective piping system, pipes of 80, 100, 125, and 150 mm diameters were selected to compute the diverse system factors.

(HP is computed from eq.16)

Required HP of pump if 80 mm pipes are used:
(Calculated TDH for 80 mm pipe = 62.73 m)

$$HP = \frac{14 \times 62.73}{75 \times 0.75 \times 0.90} = 17\ HP$$

Cost of pumping = 17 × 0.745 × 3.00 = Rs. 38.82 / hour

Required HP of pump if 100 mm pipes are used:
(Calculated TDH for 100 mm pipes = 33.97 m)

$$HP = \frac{14 \times 33.97}{75 \times 0.75 \times 0.90} = 10\ HP$$

Cost of pumping = 10 × 0.745 × 3.00 = Rs. 21.00 / hour
Annual cost = 21.00 ×8 × 25 ×12 = Rs. 53, 712

Required HP of pump if 125 mm pipes are used:
(Calculated TDH for 125 mm pipes = 24.34 m)

$$HP = \frac{14 \times 24.34}{75 \times 0.75 \times 0.90} = 7\ HP$$

Cost of pumping = 7 × 0.745 × 3.00 = Rs 15.00 / hour
Annual cost = 15.00 × 8 × 25 × 12 = Rs. 37, 598

Required HP of pump if 150 mm pipes are used:
(Calculated TDH for 150 mm pipes = 21.70 m)

$$HP = \frac{14 \times 21.70}{75 \times 0.75 \times 0.90} = 6\ HP$$

Cost of pumping = 6 × 0.45 × 3.00 = Rs.13.33 / hour
Annual cost =13.33× 8 × 25 ×12 = Rs. 31, 992

Cost difference = Rs. 5606.00

Similar to this example, pipes of various materials, such as plastic, or additional pipe diameters could be chosen to evaluate cost-effectiveness.

Cost calculations**:** The variation in recurring costs involving the above diameter pipelines is based on the following assumptions:

1. Power cost: 3.00 Rs/kwh (under revision)
2. Pump efficiency: 75%
3. Motor efficiency: 90%
4. The well is pumped intermittently for eight hrs a day, for 25 days in a month.

Table 37: Pipeline selection (Abstract)

Pipe diameter (mm)	HP of pump	Annual pumping cost	Difference in cost
80	17	Rs.91,200.00	
100	10	Rs.53,712.00	Rs.37,488.00
125	07	Rs.37,598.00	Rs.16,144.00
150	06	Rs.31,992.00	Rs. 5,606.00

The cost analysis indicates significant differences in recurring costs between 80 and 100 mm system pipes, and progressively diminishing cost differences with larger diameter pipelines. The difference in cost reduction becomes insignificant between 125 and 150 mm pipes. It would appear from the study of the friction Head computations, and the cost analysis that 100 mm pipe is the most economical and cost-effective diameter to convey 14 LPS water over a distance of 300 meters, with a static Head of 20 meters.

The capital investment (C) of the pipes and the recurring costs are equated as follows:

$$A = p \times h \times e = \text{recurring cost factor}$$

$$B = \frac{e\,(I + a)}{100} = \text{non recurring capital cost factor}$$

Where:

A = Annual power cost
p = Power consumption in kwh
h = Operating hours per annum
e = Cost per kwh
B = Annual depreciation
C = Capital investment for pump and pipe line.
I = Interest in %
a = amortization in %

The pipe diameter should be so selected as to reflect the least A + B cost factors. In cases where such cost analysis studies have been made on widely differing pipe diameters, the economy accruing from deploying intermediate diameter pipes could be determined by plotting values of A + B as abscissa, against cost as ordinate on an arithmetical graph. The economic internal diameter of the pipe would show up at that point where the resultant A + B curve transits a trough or low point.

CHAPTER XV

COMPOSITION AND PURIFICATION OF WATER

The chemical and biological characteristics of water determine its usefulness for either industry, agriculture, or home. Water may look sparkling and inert, but at the same time, it may hold in solution lethal chemicals and bacteria, making it toxic for either short term or long-term use. Establishing the quality of water, for any particular use is of scientific, economic, and hygienic consequence, and should form an integral part of any water supply mission.

The number of most important dissolved constituents in groundwater is largely limited, and the natural variations are not as large as would appear from a study of the complex mineral and organic regime through which groundwater transmits.

'TOTAL DISSOLVED SOLIDS' in a water sample includes all solid material in solution, whether ionized or not. It does not include suspended sediment, colloids, or dissolved gases. The simplest classification of water quality is based on the total concentration of dissolved solids:

Table 38: Water quality classification (ppm = parts per million)

Concentration of total dissolved solids in ppm	Classification
00 to 1000	Fresh water
1000 to 10000	Brackish water
10000 to 100000	Saline water
In excess of 100000	Brine

Table 39 lists the chemical standards prescribed for common uses of water.The safe limits are compiled from WHO international health standards. Max values differ marginally from country to country, as decided by the respective Governments.

Table 39: Maximum tolerable radical concentrations in ppm

Constituent	Drinking (max)	Household Utility		Irrigation	Food Process
		Good	Poor		
Antimony	0.05				0.05
Arsenic	0.05				0.05
Barium	1.00				1.00
Bicarbonate	500.00	150	500	200	300
Boron	20.00			0.30	
Cadmium	0.01				0.01
Calcium	200.00	40	100		80.00
Chlorides	250.00			100	300.00
Chromium	0.05				0.05
Copper	1.00	0.50	3.00		3.00
Cyanide	0.20				0.20
Fluorides	1.50				1.50
Hydrogen sulfide	1.00	0.05	2.00		0.50
Iron	0.20	0.20	0.50		0.20
Lead	0.05				0.05
Magnesium	125.00	20.0	100.0		40.0
Manganese	0.05	0.05	0.30		0.10
Nitrates	20.00				20.00
Phenol	0.001				0.001
Selenium	0.01				0.01
Silica	0.00	10.00	50.00		50.00
Silver	0.05				0.05
Sodium	200.00	100.0	300.0	300.0	300.00

Constituent	Drinking (max)	Household Utility		Irrigation	Food Process
		Good	Poor		
Sulphate	250.00	100.0	300.0	500.0	
Synthetic detergents	0.50	0.20	1.0		0.50
Uranium	20 ng/ml				
Zinc	5.00				5.00
Total solids	1500.00	300.0	2000	500.0	1000.0

Values recommended concerning possible uses of water should be applied with caution, and not with any rigidity. For example, seawater can be used for washing, cooling, recreation, and fire protection. In parts of Australia, North Africa, India, and some arid countries, people have been drinking water with dissolved solids exceeding classified limits, without suffering any injurious effects.

Broad indicators and significance of chemicals as impurities in water:

1) Water has a high dielectric constant, and consequently has the properties of an universal solvent. It can dissolve trace amounts of almost everything, even rocks. Many compounds, whether hydrated or anhydrous, dissolve in water to give electrolytic solutions of hydrated cations or anions. Even predominantly covalent compounds like alcohols, organic acids, ketones and ethers dissolve to a large extent in water.
2) Water with pH (hydrogen ion concentration) greater than 7 is alkaline, and less than 7 is acidic. Distilled water has a pH of 7 (neither alkaline nor acidic) and sea water has an average pH of 8.3 (slightly alkaline). If the water is acidic (lower than 7), lime, soda ash or sodium hydroxide is added to raise the pH. An alkaline environment is healthier than an acidic one.
3) More than 0.3 ppm of dissolved iron frequently results in appearance of reddish water, shortly after exposure, due to precipitation of hydrous iron oxide. Excessive iron will stain plumbing fixtures, stain clothes, and metallic containers.

4) Dissolved manganese will react in a manner similar to presence of iron, giving a black stain, if more than 0.1 ppm is present.
5) Persons with a sensitive taste can detect chlorides in excess of 300 ppm. Most people cannot tolerate drinking water that contains more than 300 ppm carbonate, 1500 ppm chlorides, or 200 ppm sulphates.
6) Water with excess carbonates precipitates a thin white crust of the chemical on heating and cooling. In areas where the water is "hard" (that is, containing significant dissolved calcium salts), boiling decomposes the bicarbonate ions, resulting in partial precipitation as calcium carbonate. This is the "fur" that builds up on boiling elements, in hard water areas.
7) Water having more than 500 ppm sulphates normally has a laxative effect when first used for drinking.
8) Boiler waters containing more than 50 to 100 ppm of alkalis (sodium-potassium) often causes foaming.
9) Drinking water containing fluorides in concentrations of 1.0 to 1.5 ppm is good as a prophylactic, deterring on-set of tooth decay; but, if present in excess of 1.5 ppm causes mottling of teeth with an unsightly stain, particularly in growing children. A still higher concentration causes flurosis—a debilitating disease causing severe bone deformation.
10) Groundwater drawn from very deep bore wells approximating 500 m in depth, drilled in hard rocks often access primeval juvenile waters (waters locked up in fractures and cavities during the formation of the rock itself). Dissolved solids have been concentrating in these waters over several thousands of years. Such waters are highly deleterious to health causing divergent diseases, if consumed continuously. Water being abstracted from such deep wells in hard rocks should be continuously and constantly monitored for its chemical content. Such wells, with continuous pumping tend to draw waters from still deeper zones.
11) Most animals can comfortably tolerate up to 10000 ppm of TDS without injury to themselves. They endure sodium chloride and magnesium sulphate rich waters, up to a certain limit, because of their strong digestive systems, but they should not be continuously provided with such water to drink. It is found that if the mineralization is very high, animals need a

period of 'acclimatization' before their systems become adapted to such water.

Water for irrigation:

The suitability of water for irrigation depends largely on two basic factors:

1) Salinity—i.e., total dissolved solids present at any point of time;
2) Sodium content in relation to calcium and magnesium.

When irrigation water high in sodium content reaches the soil, the clay present in the soil takes up most of the sodium. The clay in turn gives up calcium and magnesium in exchange for sodium. This is called "Base Exchange." This exchange of ions alters the physical and chemical characteristics of the soil. Clay that carries excess of calcium (Ca) or magnesium (Mg) tills easily and possesses good permeability. The same clay, when it acquires sodium from the irrigation waters, becomes sticky when wet, and looses permeability. It shrinks into hard clods when dry, which are difficult to break up by cultivation.

High concentration of sodium increases alkalinity in the soil, in which little or no vegetation would grow. However, if the irrigation water contains calcium and magnesium ions in quantities that exceed, or equal the sodium content, all these salts will be retained on the clay particles. Then the soil will maintain good 'tillability', as also permeability, necessary for good growth. Such waters are good for irrigation, though the total salinity may appear high. Hence, it is important to remember while evaluating the chemical quality of the irrigation water, that absolute values of sodium are not as significant as the ratio of sodium to calcium and magnesium.

$$\text{Percent alkalis} = \frac{\text{epm sodium and potassium}}{\text{epm sodium + epm potassium + epm magnesium}} \times 100$$

Percentage of alkalis (sodium and potassium) by weight is obtained by converting ppm of sodium, potassium, calcium and magnesium to equivalents per million(epm) by multiplying by the appropriate factor in column 2 of the conversion Table 46.

Sodium percentage is a reliable indicator of the suitability of water for irrigation. This percentage is actually the ratio of the sodium ions to the total sodium, calcium and magnesium ions. If the sodium percentage exceeds 50%, it implies that the sodium content exceeds that of calcium and magnesium, and hence this excess sodium becomes harmful to the soil, and in turn to cultivation. A sodium percentage in the water exceeding 50 percent of the total Bases is potentially harmful for cultivation. This is called 'sodium hazard'.

'SAR' the short form for 'sodium adsorption ratio' indicates the quantum of deleterious effect of the irrigation waters carrying excess sodium. It is calculated from the following equation, the concentration of the ions being expressed in epm:

$$SAR = \frac{26Na^{+}}{\sqrt{Ca^{++} + Mg^{++}}}$$

SAR values of 18 or more indicate potential high sodium hazard; values between 10 and 18 moderate sodium hazard; while values below 10 indicate 'nil' sodium hazard.

Irrigation waters have been classified into Class 1, Class 2, and class 3 categories (C.S.Scolfield) as given in table 40:

Table 40: Classification of irrigation waters in ppm

Classification	Total dissolved solids (TDS)	Chlorides	Boron	Percent alkalis
Class 1	0 to 700	0 to 150	0 to 0.5	< 60%
Class 2	700 to 2000	150 to 500	0.5 to 2.5	60 to 75%
Class 3	Excess of 2000	Excess of 500	Excess of 2.5	Excess of 75%

Class 1 waters are regarded as entirely safe for irrigation under common conditions of climate and soil even for sensitive crops; class 2 waters are of intermediate value, being safe under certain conditions only for specific crops. Such waters need careful analysis. Class 3 waters with concentrations of one or more constituents in excess, are unsafe for irrigation of majority of crops. In addition to water analysis, other pertinent factors such as climatic conditions, extent of irrigation relative to rainfall, nature of soil, specificity of the crop is also considered and integrated with the chemical quality of the water to be used for irrigation. These factors are of particular importance for groundwater usage for irrigation. Boron is an element, which is essential for plant growth, but is injurious in large quantities.

While interpreting chemical analyses data, the amount of each constituent is converted from 'parts per million' (ppm) to 'equivalents per million' (epm) by dividing ppm by the chemical combining weight of each ion. Frequently chemical analyses data is provided in terms of epm instead of ppm. Expressed in this way, equal concentrations of different ions are equivalent to their tendency to form possible chemical combinations. Multiplying ppm by the following conversion factors provide the concentration of each of the ions in epm. Chemical analyses data is also given in terms of ppm with reference to $CaCo_3$. In either presentation, it is easier to convert the data to simple ppm to assess the suitability of the water. Multiplying parts per million by the following conversion factors provide the concentration of each of the ions in equivalents per million. Hardness, alkalinity and acidity of water is usually reported in terms of calcium carbonate ($CaCO_3$):

Calcium (Ca^{++}) 0.04990: Magnesium (Mg^{++}) 0.08224:
Sodium (Na^{+}) 0.04348

Table 41: To convert epm to ppm; ppm to epm and ppm to ppm $CaCO_3$ *(abstracted from WHO standards)*

Constituent	epm to ppm	ppm to epm	ppm to epm $CaCo_3$
Calcium, Ca	20.04	0.04991	2.4970
Iron, Fe	27.92	0.03582	1.7923
Magnesium, Mg	12.16	0.08224	4.1151
Potassium, K	39.10	0.02558	1.2798

Constituent	epm to ppm	ppm to epm	ppm to epm $CaCo_3$
Sodium, Na	23.00	0.04348	1.1756
Bicarbonate, HCO_3	61.01	0.01639	0.8202
Carbonate, CO_3	30.00	0.03333	1.6680
Chloride, Cl	35.46	0.02820	1.4112
Hydroxide, OH	17.01	0.05879	2.9263
Nitrate, NO_3	62.01	0.01613	0.8070
Phosphate, PO_4	31.67	0.03158	1.5800
Sulphate, SO_4	48.04	0.02082	1.0416
Calcium bicarbonate, $Ca(HCO_3)_2$	81.04	0.01234	0.6174
Calcium carbonate, $CaCO_3$	50.04	0.01998	1.0000
Calcium chloride, $CaCl_2$	55.50	0.01802	0.9016
Calcium hydroxide	37.05	0.02699	1.3506
Calcium sulphate, $CaSO_4$	68.07	0.01469	0.7351
Ferrous bicarbonate, $Fe(HCO_3)_2$	88.93	0.01124	0.5627
Ferrous carbonate, $FeCO_3$	57.92	0.01727	0.8640
Ferrous sulphate, $FeSO_4$	75.96	0.01367	0.6588
Magnesium bicarbonate, $Mg(HCO_3)_2$	73.17	0.01367	0.6389
Magnesium carbonate, $MgCO_3$	42.16	0.02372	1.1869
Magnesium chloride, $MgCl_2$	47.62	0.02100	1.0508
Magnesiumhydroxide, $MgOH_2$	29.17	0.03428	1.7155
Magnesium sulphate, $MgSO_4$	60.20	0.01661	0.6312
Potassium chloride, KCl	74.56	0.01341	0.6711
Sodium bicarbonate, $NaHCO_3$	84.01	0.01191	0.5956
Sodium chloride, NaCl	58.46	0.01711	0.8560
Sodium hydroxide, NaOH	40.01	0.02449	1.2507

Constituent	epm to ppm	ppm to epm	ppm to epm $CaCo_3$
Sodium nitrate, $NaNO_3$	85.01	0.01176	0.5886
Sodium phosphate, Na_3PO_4	54.67	0.01829	0.9153
Sodium sulphate, Na_2SO_4	71.04	0.01408	0.7044

1. For converting equivalents per million—epm—to parts per million—ppm—multiply epm by factors in column 1.
2. For converting parts per million—ppm—to equivalents per million—multiply ppm by factor in column 2.
3. For converting parts per million—ppm—to parts per million of $CaCo_3$—ppm $CaCo_3$—multiply computed figures in column 1 by factors in column 3.

Many laboratories report hardness, alkalinity and acidity in terms of calcium carbonate. ppm is also defined as, or is equivalent to one milligram of calcium carbonate ($CaCo_3$) per liter of water.

Water softening:

Every household and every industry uses water, and none of it is pure. Water carries salts in the form of dissolved solids, many of them essential for health and well-being, but the critical factor is the quantity of the chemical held (Table 39). Pure water, such as distilled water is greatly injurious to health.

The dissolved calcium and magnesium salts (+ iron and aluminum, if present in large quantities) cause hardness in water. The bi-carbonates of calcium and magnesium cause carbonate hardness called temporary hardness, while the sulphates and chlorides cause non-carbonate hardness called permanent hardness. In hard waters, soap produces a white scum like precipitate, instead of lather, which retards cleaning. Hardness in water depends on the amount of Ca and Mg held in solution and is usually judged by the quantity of soap required to produce lather. These salts also precipitate in the pipes and connecting hoses, often clogging the conduits.

'Par se' hard water consumption does not cause any adverse health effects in either humans or animals. In fact, its mineral content is a contributory factor in sustaining the mineral requirements of the body.

Hard water is commonly softened in large quantities by either the lime-soda-ash or the zeolite process. In the lime-soda-ash process, lime (calcium hydroxide) is added to the water to precipitate the calcium carbonate and magnesium salts. Soda ash (sodium carbonate) is added to precipitate the calcium sulphate and calcium chloride. These insoluble compounds are then evacuated from the water by sedimentation and filtration.

In the zeolite process, the hard water is passed through natural or synthetic zeolite, which has the property of exchanging its sodium (sodium is a base integer of zeolites) for calcium and magnesium in the water. After softening a quantity of water, the zeolite is 'regenerated' with the common salt (sodium chloride) solution to replace the sodium taken by the hard water. A distinct type of zeolite has been developed which has the unique property of eliminating most of the dissolved minerals in water, the finished water approximating distilled water under certain conditions.

Large quantities of water, such as required for municipal or industrial supplies is softened by either of the above methods; but the lime-soda-ash process is probably more widely used. Water supplied to households and other establishments, through pipes by the City Corporations have been pre-treated and purified for drinking. These waters do not require any further treatment, except normal filtration through a membrane and a carbon filtrate. However, softening by the zeolite process becomes necessary when soft water of zero hardness is required as an integral input, such as in certain industries. Water softeners chemically replace the calcium and magnesium with sodium. Sodium does not react with soap, produces normal lather, and does not form scales inside pipes. However, from a health perspective, Ca and Mg are better and healthier for our body systems than sodium.

Flocculation:

Flocculation is a process that clarifies the water. (Clarifying means removing any turbidity or colour so that the water is clear and colorless.) Clarification is done by causing a precipitate to form in the water that can be removed using simple physical methods. Initially the precipitate forms as very small particles but as the water is gently stirred, these particles stick together to form bigger particles. In this way, the coagulated precipitate takes most of the suspended matter out of the water and is then filtered off, generally by passing the mixture through a coarse sand filter or sometimes through a mixture of sand and granulated anthracite (high carbon and low volatile coal). Coagulants / flocculating agents that may be used include aluminum, hydroxide (alum) and iron hydroxide.

Water purification:

Water purification is the process of removing undesirable chemicals, materials, and biological contaminants from contaminated water. There are simple and complex procedures to purify water. The procedure employed take into account the volume of pure water required, and the end use of the purified water. Most water intended for drinking is purified for human consumption, but water purification may also be designed for a variety of other purposes, including meeting the requirements of medical, pharmacology, chemical and industrial applications. The standards for drinking water quality are typically set by governments or by international standards. It is not possible to advise whether water is of an appropriate quality by simple visual examination. Minimal procedures such as boiling or the use of a household activated carbon filter are not sufficient for treating all the possible contaminants that may be present in water from an unknown source. Chemical analysis is the only way to obtain the information necessary for deciding on the appropriate method of purification. Surface water from rivers, canals, and reservoirs will have a significant bacterial load, suspended solids, and a variety of dissolved constituents.

Water purification techniques:

1) **Boiling:**
   Water is boiled for one minute to inactivate or kill micro-organisms that normally live in water at room temperature. With the exception of calcium, boiling does not remove solutes of higher boiling point than water and in fact increases their concentration (due to some water being lost as vapor). Boiling does not leave a residual disinfectant in the water. Therefore, water that has been boiled and then stored for any length of time may have acquired new pathogens.

2) **Granular Activated Carbon filtering:**
Granular Activated Carbon is a form of activated carbon with a high surface area, adsorbs many compounds including many toxic compounds. Many household water filters use activated carbon filters to purify the water. Household filters for drinking water sometimes contain silver as metallic silver nanoparticle. If water is held in the carbon block for longer period, microorganisms can grow inside which results in fouling and contamination. Silver nanoparticles are excellent anti-bacterial material and they can decompose toxic halo-organic compounds such as pesticides into non-toxic organic products.

3) **Disinfection:**
Chlorine disinfection, the most common disinfection method involves some form of chlorine or its compounds such as chloromines or chlorine di oxide. Chlorine is a strong oxidant that rapidly kills many harmful microorganisms. Because chlorine is a toxic gas, there is a danger of its release associated with its use. This problem is avoided by the use of sodium hypochlorite which is a relatively inexpensive solution that releases free chlorine when dissolved in water. A solid form, calcium hypochlorite releases chlorine on contact with water.

4) **Distillation:**
Distillation involves boiling the water to produce water vapor. The vapor as it contacts a cool surface, it condenses as a liquid. Because the solutes are not normally vaporized, they remain in the boiling solution. Even distillation does not completely purify water, because of contaminants with similar boiling points and droplets of unvaporised liquid carried with the steam. However, 99.9% pure water can be obtained by distillation. Distilled water is not fit for drinking.

5) **Reverse osmosis:**
Normal osmosis is a natural life process, without which life as we know today would not exist. Osmosis is the movement of solvent molecules through a selectively permeable membrane into a region of higher solute concentration, aiming to equalize the solute concentrations on the two sides. It may also be used to describe a physical process in which any solvent moves, without input of energy

across a semi permeable membrane (permeable to the solvent, but not the solute) separating two solutions of different concentrations. Net movement of solvent is from the less concentrated (hypotonic) to the more concentrated (hyperton) solution, which tends to reduce the difference in concentrations. This is the reason why ship wrecked sailors, at extreme thirst conditions drink seawater and death follows within minutes. Sea water being denser with high concentrations of salt tend to draw in the less dense water in the body using the stomach membrane as a semi-permeable membrane through which the water in the body gushes in to equalize the solute concentration within and outside the stomach. Death occurs because of dehydration.

Reverse osmosis is essentially applied as a desalinization process. It is most commonly known for its use in eliminating salts from seawater. It is a separation process that uses pressure to force a solvent through a semi-permeable membrane that retains the solute on one side and allows the pure solvent to pass to the other side. More formally, it is the process of forcing a solvent from a region of high solute concentration through a membrane to a region of low solute concentration by applying a pressure in excess of the osmotic pressure. Reverse osmosis is theoretically the most thorough method of large-scale water purification available, although perfect semi-permeable membranes are difficult to create. Unless membranes are well maintained algae and other life forms, in course of time colonize the membranes.

6) Forward osmosis:

Osmosis may be used directly to achieve separation of water from a "feed" solution containing unwanted solutes. A "draw" solution of higher osmotic pressure than the feed solution is used to induce a net flow of water through a semi-permeable membrane, such that the feed solution becomes concentrated as the draw solution becomes dilute. The diluted draw solution may then be used directly (as with an ingestible solute like glucose), or sent to a secondary separation process for the removal of the draw solute. This secondary separation can be more efficient than a reverse osmosis process would be alone, depending on the draw solute used and the feed water treated. Forward osmosis is an area of ongoing research, focusing on applications in desalination, water purification, water treatment, food processing etc.

Distillation removes all minerals from water, and the membrane methods of reverse osmosis and nanofiltration remove most to all minerals. This results in demineralized water which is not considered ideal potable water. Desalination processes are known to increase the risk of bacterial contamination in drinking water. The World Health Organization has investigated the health effects of demineralized water since 1980. Experiments in humans found that demineralized water increased diuresis and the elimination of electrolytes, with decreased blood serum potassium concentration. Reverse osmosis, along with other radicals eliminates magnesium, calcium, and minerals that are essential solutes in drinking water, which help to protect the human metabolic system against nutritional deficiency. Recommendations for magnesium have been put at a minimum of 10 mg/L with 20-30 mg/L as optimum; and for calcium a 20 mg/L minimum and a 40-80 mg/L as optimum. At water hardness above these limits, higher incidence of gallstones, kidney stones, urinary stones, arthrosis, and arthropathies have been observed.

Manufacturers of home water distillers claim the opposite—that minerals in water are the cause of many diseases, and that most beneficial minerals come from food, not water. They quote the American Medical Association as saying "The body's need for minerals is largely met through foods, not drinking water." The WHO report agrees that "drinking water, with some rare exceptions, is not the major source of essential elements for humans" and is "not the major source of our calcium and magnesium intake," yet states that demineralized water is harmful anyway. "Additional evidence comes from animal experiments and clinical observations in several countries. Animals given zinc or magnesium dosed in their drinking water had a significantly higher concentration of these elements in the serum, than animals given the same elements in much higher amounts with food and provided with low-mineral water to drink."

7. Atmospheric water generation is a new concept to provide high quality drinking water by extracting water from the air by cooling the air to condense the water vapor. This is an expensive technology and can be installed only in very dry areas.

8. Fog condensation collects water from the atmosphere in areas with significant dry seasons and in areas, which experience fog even when there is little rain.

Purity of groundwater:

The water emerging from some deep underground source may have fallen as rain many tens, hundreds, thousands of years ago. Groundwater being constantly mobile, would have traveled inestimable distances and traversed various underground formations. Soil and rock layers, through which the groundwater travels, filter naturally the ground water to a high degree of purity. Concomitantly, it also takes into insipient solution some of the mineral salts. Such water may emerge as springs, artesian wells or may be abstracted through boreholes or wells. Deep ground water is generally of very high bacteriological quality (i.e., pathogenic bacteria or the pathogenic protozoa are typically absent), but the water typically is rich in dissolved solids, especially carbonates and sulfates of calcium and magnesium. Depending on the formations through which the water has traversed, other ions may also be present including chlorides and bicarbonates. It may be necessary to reduce the iron or manganese content to make water pleasant for drinking. Disinfecting the water may also be a necessity. Where groundwater recharge is practiced (a process in which river water is injected into an aquifer to store the water so that it is available in times of drought), the groundwater should be considered equivalent to lowland surface waters for treatment purposes.

An effective, and economical method of disinfecting wells, tube wells, tanks, pools and such water holding areas is by the use of calcium hypochlorite (chlorinated lime) containing approximately 25% available chlorine. The chemical cannot be stored for long. A fresh supply is essential, since the chemical deteriorates on prolonged exposure. If high-test calcium hypochlorite (H-T-H), which contains about 70% available chlorine is used, the quantity required will be about one-third the amount of chlorinated lime.

Table 42 indicates the quantity of chlorinated lime recommended for disinfecting wells and tanks of differing capacities.

Table 42: Recommended quantities of chlorinated lime for disinfecting wells/tanks of various capacities *(Johnson)*

Well/Tank Capacity (litres)	Chlorinated lime to be added (gms)	Water to be added for preparing solution (litres)
200	42	19
350	85	19
750	170	19
1100	255	19
1500	340	19
1900	425	19
3750	850	38
7575	1700	55
11350	2550	75

Procedure: To the recommended amount of chlorinated lime, gradually add small incremental quantities of water, and stir until a smooth paste devoid of lumps is formed. To this paste add recommended quantity of water, stir for 10 to 15 minutes, and allow the solution to settle. After the solution fractionates and settles, the clearer liquid containing chlorine is utilized, while the inert material, containing largely of lime that has settled at the bottom is discarded. Strong chlorine solution corrodes metal; hence, metal containers should be avoided.

For disinfecting small quantities of water, the required quantity of chlorinated lime could be measured with a spoon; one moderately heaped teaspoon of chlorinated lime weighs approximately 30 gms. For example, to disinfect 200 lts of water, 1½ tea spoons or 1 spoon of chlorinated lime stirred in 20 lts of water would provide the required concentrate.

Disinfecting tube wells:

i) When a tube well or a bore well is tested for its yield, the test pump should be run until the water is clear, and free from turbidity;

ii) After removing the test pump, and before installing the selected production assembly, the well should be disinfected, by pouring the recommended amount of chlorinated lime into the well slowly to mix with the water in the well.
iii) Wash the pump with chlorinated lime and the drop pipe as they are lowered into the well.
iv) When the pump is in position, run the pump intermittently for very short periods (about 30 seconds) to further mix the solution and as to wash the interior of the pump.
v) After the chlorine solution has been mixed, and has circulated through the pumping system, shut down the pump, and allow the solution to remain in the well for about 10 hours.
vi) Lastly, the well should be cleaned by pumping, until the water is free from the odor of chlorine.

In case of deep wells with high static water levels, it may be desirable to place the chlorinated lime in a short length of perforated pipe, capped at both ends, and lowering and gradually raising the pipe through the water to the surface. Disinfecting a well will eliminate the bacteria present in the well, and on the pumping equipment. However, if there is any source of external contamination, the procedure will be effective temporarily, limited to a single application.

Common contaminants of groundwater:

Most of the ground waters, by reason of their travel through formations, which act as natural filters and absorbents, are suitable for consumption; but concomitantly and occasionally, these waters dissolve salts which, in certain excessive solutions may be injurious to health. Before drinking untreated groundwater, it is advisable to test ppm limits of its radicals in a testing laboratory.

Flouride:

Most injurious are the fluoride and arsenic containing waters. Fluoride incidence beyond desirable amounts (0.6 to 1.5 mg/l) in water is a serious problem in many parts of the world. Around 200 million people from 25 nations are exposed to grave health risks caused by high amounts of fluoride in groundwater (Ayoob and Gupta 2006). In

India too, there has been an increase in incidence of dental and skeletal fluorosis with about 62 million people at risk (Andezhath et al.1999) due to high fluoride concentration in drinking water. Dental fluorosis is endemic in parts of 14 states and 150,000 villages in India with the problem most pronounced in the states of Andhra Pradesh, Bihar, Gujarat, Madhya Pradesh, Punjab, Rajasthan, Tamil Nadu and Uttar Pradesh. (Brinda et al, 2011))

Numerous methods have been developed by multifarious Organizations all over the world to eliminate or dilute this menace of fluoride occurrence in groundwater. Every method has certain economic, social, quantum and rapidity advantage and disadvantage.

IGRAC (International groundwater resources assessment centre), an initiative of the UNESCO World Meteorological Organization (WMO); Partners in water, Government of Netherlands; and TNO Built Environment and Geosciences, hold and publish literature on the various methods in vogue for removal of fluorides. Their contact details are: IGRAC, P.O.Box 80015, 3508 TA Utrecht, The Netherlands. e-mail: info@igrac. nl: internet > www.igrac.nl.

Amongst the economical methods available to eliminate fluoride are:

1) Adsorption by clay: Clay is fine-textured, plastic when moist, retains its shape when dried and sinters hard when fired. Both clay powder and fired clay are capable of sorption of fluoride as well as other pollutants from water. Process is commonly in use in Sri Lanka.
2) Activated alumina: Activated alumina is aluminium oxide (Al_2O_3) grains prepared to have an adsorptive surface. When the polluted water passes through a packed column of activated alumina, pollutants and other components in the water are adsorbed onto the surface of the grains.
3) Bone Charcoal: Bone charcoal is a blackish, porous, granular material. The major components of bone charcoal are calcium phosphate 57-80 per cent, calcium carbonate 6-10 per cent, and activated carbon 7-10 per cent. In contact with water, the bone charcoal is able, to a limited extent, to absorb a wide range of pollutants such as color, taste, and odor and has the specific ability to take up fluoride from water.

4) Nalgonda process: Process is based on aluminum sulphate Coagulation,-flocculation, and sedimentation where the dosage is designed to ensure fluoride removal from the water. Two chemicals, alum (aluminum sulphate or potassium aluminum sulphate) and lime (calcium oxide) are rapidly mixed with the fluoride-contaminated water. Induced by subsequent gentle stirring, by a hand operated mechanical devise, wispy wool like gossamer threads of aluminum hydroxide precipitates in the solution. These wisps carry most of the dissolved fluoride and are removed after being allowed to settle. The clean water is moved up, and is tapped out at the top of the cylinder. The quantity of chemicals being added are regulated proportionate to the volume of water under filtration. The Nalgonda process to eliminate fluoride was adapted, and further developed in India by the National Environmental Engineering Research Institute (NEERI) and has been modified suitably for use at both the community and household levels. The Nalgonda technique has been introduced in many countries, including India, Kenya, Senegal and Tanzania amongst others, where these units are working successfully both at community and household levels.

In view of acute shortage of water, and the deleterious role of fluoride in ground water in many parts of the world, continual research in developing simple inexpensive set-ups, suitable in third world households are essential.

Arsenic:

Arsenic is another dangerous radical that contaminates groundwater in many parts of the world. Arsenic is a high potency poison, is carcinogenic and in excess of 10 μ/l is suspected to cause onset of diseases like diabetes and renal failure. Arsenic contamination of groundwater is often due to naturally occurring high concentrations of arsenic in deeper levels of groundwater occurrence. Arsenic contamination of ground water is found in many countries throughout the world, including the USA, Thailand, Taiwan, Mainland China, Nepal, Argentina, and Chili. There are also many locations in the United States where the groundwater contains arsenic concentrations in excess of the Environmental protection Agency standard of 10 parts per billion. According to a recent film funded by the US Superfund, "in small doses". millions of private wells have unknown arsenic levels, and in some areas of the US, over 20 %.(data ref from Wikipedia).

Arsenic contamination in groundwater in the Ganga—Brahmaputra—Imphal fluvial plains in India, and Padma—Meghna fluvial plains in Bangladesh have been widely reported. Arsenic contamination of the groundwater in Bangladesh is a serious problem affecting vast sectors of its population. Throughout Bangladesh, as tube wells are tested for concentrations of arsenic, those that are found to have arsenic concentrations over the amount considered safe, are painted red to warn residents that the water is not safe to drink.

In India, seven states namely—West Bengal, Jharkhand, Bihar, Uttar Pradesh, Assam, Manipur and Rajnandgaon village in Chhattisgarh state have so far been reported to be affected by Arsenic contamination in groundwater which contains arsenic above the permissible limit of 10 μg/l.

In the Ganges Delta, the affected wells are typically more than 20 meters and less than 100 meters deep. It has been theorized that Groundwater closer to the surface typically has spent a shorter time in the ground, therefore likely absorbing a lesser concentration of arsenic; while water at depths deeper than 100 m is exposed to older formations, which are leached of arsenic. There is no corroborative study on this supposition, but since a nexus between depth and incidence of arsenic appears to be true, investigation of arsenic occurrence in ground waters at various depths should form a part of the study.

There are various methods of elimination of arsenic from water, each of them with a certain limitations. (Ghosh N.C.)

Principal amongst them are:

Coagulation and filtration(also known as flocculation): Ferric oxide has a high affinity for adsorbing certain dissolved radicals that includes arsenic. Iron coagulants remove arsenic from the water by precipitation and adsorption. However, the difficulty with this type of filtration system is that the filters get clogged within two to three months. The toxic arsenic sludge is disposed of by concrete stabilization, but there is no assurance that they will not leach out in course of time.

Activated alumina is an adsorbent that effectively removes arsenic. Activated alumina columns connected to shallow tube wells in India and Bangladesh have been successfully deployed to remove arsenic from groundwater. Long-term column performance has been possible through the efforts of community-elected water committees that collect a local water tax for funding operations and maintenance. It has also been used to remove undesirably high concentrations of fluoride (Sudipta Sarkara 2005).

Both Reverse osmosis and electro-dialysis can remove arsenic with a net ionic charge. Arsenic oxide, As_2O_3, is a common form of arsenic in groundwater that is soluble, but has no net charge. Some utilities presently use one of these methods to reduce total dissolved solids. A problem with both methods is the production of high-salinity wastewater, or concentrate, which then must be safely disposed of.

Subterranean Arsenic Removal (SAR) Technology: In subterranean arsenic removal (SAR), (Soumyadeep Mukherjee) aerated groundwater is recharged back into the aquifer to create an oxidation zone which can trap iron and arsenic on the soil particles through adsorption process. The oxidation zone created by aerated water boosts the activity of the arsenic-oxidizing microorganisms, which can oxidize arsenic. No chemicals are used and almost no sludge is produced during operational stage since iron and arsenic compounds are rendered inactive in the aquifer itself. Six such SAR plants, funded by the World Bank and constructed by Ramakrishna Vivekananda Mission, Barrackpore & Queen's University Belfast, UK are operating in West Bengal. Each plant has been delivering more than 3,000 liters of arsenic and iron-free water daily to the rural community.

Currently, large-scale SAR plants are being installed in USA, Malaysia, Cambodia, and Vietnam.

The Hungarian solution:

Hungarian engineer László Schremmer has recently discovered that by the use of chaff-based filters it is possible to reduce the arsenic content of water to 3 microgram/liter. Chaff is the dry scaly protective casings of the seeds of cereal grains or similar fine, dry, scaly plant material or finely chopped straw.

APPENDIX 1

TROUBLE SHOOTING A TUBE OR BOREWELL

The well at the time of construction should be provided with the following permanent fixtures that act as diagnostic tools to analyze any problems that may arise during the service life of the well:

1. A pressure gauge and a throttle valve along the discharging arm.
2. A water level indicator, or preferably 12mm or ½ inch flexible tubing (called an airline) attached to the eductor pipe and inserted along with the pump into the well down to the pump level. The line should be corked at the top, to prevent any material or insects falling into it. This tubing, in preference to a water level indicator, will facilitate insertion of a steel tape (smeared with ordinary chalk for easy reading of the wetted part of the tape) to check water levels. A water level indicator is analogous to a thermometer used for preliminary diagnosis of onset of ailments.
3. A flow meter in low yielding wells, or an appropriate devise to check output in high yielding wells.

It is possible to estimate the yield by constructing a 'Table' indicating the output pressure against various volumes of flow. With such a Table 'pressure' can be correlated to yield, and can be estimated directly, irrespective of the closure status of the throttle valve.

Wells failing to yield water during drilling:

In hard rock areas, a borehole, in spite of being drilled at a favorable location, to a sufficient depth may fail to yield water during drilling for a number of reasons.

Probable causes are:

1. Excessive lengths of blank casing inserted into borehole effectively shuts off capillary water held in the vadose zone, and groundwater circulating under pressure in the indistinct boundary between the overlying soils and the weathered and fresh rocks below. Blank casing should be limited to the upper loose soil. Weathered or disintegrated rock zones should be left open, without any casing. General depth of blank casing is 10 m. In case loose formations occur to depths greater than 10 m, the casing below this depth should be slotted or perforated;
2. Drilling under excessive pressures;
3. Insufficient well depth;
4. In case a borehole fails to yield water during drilling, stimulation procedures may help.
5. A borehole may tend to collapse during drilling, while encountering loose formations or fracture zones.

If the collapse occurs:

i) above 15 m depth, blank case the well;
ii) between 15 and 25 m depth, with groundwater exiting the well, install slotted casing down to 25 m depth;
iii) below 30 m depth, it is preferable to explore for an alternative location;
iv) If the hole collapses at greater depths, while the hole is yielding water, terminate the drilling and develop the well.

Sometimes it would be necessary to continue drilling through the fracture zones, after flushing out the collapse. Alternatively, drill a reasonable depth of about 10 m through the fracture zone, pullout the drill string, allow the loose zone to collapse, grout the zone with quick setting cement, and drill through the grouted zone.

Failure of an active well:

A functioning bore well may fail for a number of reasons, least of the reasons being lack of water in the well. Before a well fails, signs portending such failure will appear; which if recognized and diagnosed on time, failure can be preempted by adopting necessary corrective measures. Critical diagnostic signs for an impending failure are:

a) Fluctuations in well yield, or spurting of water from the discharge pipe;
b) Change in static and pumping water levels in the well;
c) Pumped water carrying inclusions such as soil or any metal particles.

To gain this data without much difficulty, the well should have been equipped with the fixtures outlined in Para 1 above, viz., throttle valve, pressure gauge and air-line to measure water level.

Drilling a second well in close proximity to a pumping well, and operating the wells simultaneously will result in drawdown interference between the wells causing the shallower well to cease yielding water early. Optimum average distance between two low out-put wells is 30 meters.

Over-pumping:

Foremost reason for a majority of the well failures is over pumping which results from i) installing a pump without a well test, ii) installing pumps with capacities exceeding the groundwater inflow into the well. Over-pumping leads in course of time, to severe lowering of the water levels resulting in pumps breaking suction and cessation of output. Continued over pumping, without any remedial measures will cause a change in the groundwater regime (reversal of pressure differentials, which cause movement of groundwater). In low yielding wells, this will lead to a permanent damage to the well location.

Remedy:At the first sign of reduced yield, control output with partial closing of the throttle valve to maintain a constant pressure on the gauge.

Declining water levels:

Many times the decline in water level is seasonal. Typically, water levels are higher post monsoon and lower in summer. Extended dry periods can also influence water levels, especially in wells tapping shallow aquifers. Pumping water level should be at least 6 m above the NPSH value of the pump.

Remedy: Check water level and lower the pump, while reducing output. Withdrawal of water should be reduced in summer, before the water levels start dropping.

Reduced output:

Pump entry filters may be clogged or encrusted.

Remedy: Pull out the pump and clean the water entry portals. A well may even temporarily cease to yield water. This may occur as a result of interference caused by an adjacent well located in close proximity, which may be pumping in tandem. When two closely spaced wells pump concurrently, the depression cones surrounding and moving out from each of the wells may in course of time, overlap each other. When this happens, each of the cones will draw the contiguous cone down, virtually doubling the normal drawdown levels. Minimum safe distance for shallow, low yielding wells drawing water from unconfined water table aquifer is 30 m, and for deep high yielding wells minimum distance should be 100 m.

Remedy: Remedy lies in not operating both the wells simultaneously. Fix alternate timings for each of the wells.

Polluted water:

Water from a normally yielding well may abruptly or during a course of time change color or taste or odor. Such changes usually indicate the groundwater being polluted from a nearby or distant source such as an industrial unit. A chemical analysis will typically pin point the polluting source.

Submersible pump failure:

Major reason for a submersible pump failure is running the pump below its NPSH (Net Positive Suction Head) or NRIH (Net required inlet Head) rating, which is equivalent to running it dry. These pumps are water-cooled machines; hence, it is important to ensure that the pump stays submerged below the pumping level. The rated Head of water above the pump's intake may range up to 7 meters. The pumps should have an appropriate capacitor, a star delta starter, a single phasing preventer, a low water level cut-off, an over-load protector and a volt-meter. These ancillaries cost only a fraction of the cost of the pump, but will ensure the safety of the system. Pumps equipped with these ancillaries rarely fail. Other reasons for a submersible pump failure are worn-out or clogged impellers, clogged intakes and a stuck check valve.

A damaged eductor or a loose welded joint will result in the well yielding less water, though the pump may be running to its full load.

A pump failure can also be pre-empted by monitoring water levels and ensuring proper functioning of the pilot instruments in the control box.

An air lock in the system may stop the pump from pumping water. Start and stop the pump repeatedly and check whether the pump resumes pumping.

Repairs to a nearby electrical main, or the supply transformer may lead to invert wiring, viz, where positive and negative leads may become reversed. In such a case, this polarity reversal will also impinge on the power feed to the pump. In such an event, the impellers will rotate in the opposite direction. The pump, though running to full load, will yield only trickle of water. An electrician can check the positive and negative leads and reverse them to normalize the pump function.

Insufficient yield from Jet pumps:

The injector nozzle and venturi tube size of the jet pump must be selected according to the pumping depth; the distance the pump is offset from the well; and the size of the pump. All pump manufacturing Companies provide charts for proper pump and injector selection. Since, in this type of pumping system, water is being continuously circulated, it is important that the proper sizes of suction and delivery pipes be used to maximize efficiency.

A well operated by a jet pump may fail to yield water under following circumstances:

1. Jet pumps will lose their prime if the water level in the well drops below the foot valve. This can be prevented, in a low yielding well, by using an automatic pressure regulator, and a weak well tailpipe below the injector. This tailpipe will automatically regulate the pumping rate to the well capacity.
2. The mechanical seal on the pump shaft is water lubricated and water cooled, so the pump should not be run without the pump casing being filled with water.
3. If the foot valve is stuck or clogged, the pump will lose its prime. In shallow wells, usually jerking the pipelines with a crow-bar or a similar implement to lift and drop the pipelines a few times will generally make the valve drop back into position. If this fails, it would become necessary to pull out the pipelines and reset or replace the foot valve.
4. Consequent to excess pumping, insufficient water may be returning to the well, causing the system to break suction. It is critical to maintain the pressure setting recommended by the manufacturer to ensure that optimum quantum of water is under circulation.

APPENDIX 2

MISCELLANEOUS READINGS IN GROUNDWATER

Conservation of water:

Groundwater, though not easily discernable, is the major natural resource, which has cradled and sustained civilizations over vast areas of the world. Groundwater occurs in far greater quantities than the surface water. However, the effort and expenditure imparted to develop groundwater, as a source of water is only a fraction of that expended to enlarge on surface water sources. This is a consequence of the apparent invisibility of the groundwater, relative to surface water.

Over the years, there has been considerable effort for development of groundwater as a sustainable resource, both by public and private endeavor. This has resulted in greater utility, and enhanced dependability on groundwater. However, developmental efforts suffer from the lacunae of deficiency in meaningful planning with long-term conservation and sustainability in view. Groundwater is a renewable resource. However, in certain areas, particularly in urban locations, groundwater withdrawal has reached a significant over-draft stage, where its exploitation is exceeding availability. Frequently, the exploitation of groundwater is approaching a critical state where it is surpassing the annual replenishment by rainfall. In such cases, groundwater regime may undergo changes transforming the respective belt to a permanently dry area. Such a situation is generally preceded by declining groundwater levels, drying up of shallow wells, intrusion of saline water far into the inland, increase in costs of lifting water and gradually affecting the socio-economic living conditions.

Ministry of water resources, Government of India, in their wide ranging surveys, based on annual rainfall and infiltration coefficients, have estimated 430 b cum (billion cubic meters: one cu m = 1000 litres or 220.083 imperial gallons) as the total annual replenishment of groundwater resource of the country. Groundwater draft is estimated at 230 b cum, meaning 50% exploitation of the available resources. Groundwater resources available for irrigation are approximately 38.284 b cum, against which draft is 10.65 b cum leaving a balance of 27.635 b cum for exploitation. Present level of groundwater development is less than 28%. However, this exploitation is highly uneven with some areas being over exploited, utilizing groundwater in excess of annual replenishment leading to drying up of the reserves. This is particularly prevalent in rapidly spreading urbanized areas on city's edges. Greatest challenge today is to balance the groundwater draft between the over exploited and under exploited regions.

Jacques Diouf, Director General, UN Food and Agricultural Organization (FAO) while submitting the FAO Report titled "State of the World's land and water resources for food and agriculture" for the year 2011 has drawn attention to the fact that " in the last 50 years, a significant increase in food production, combined with demographic pressure and unsustainable agricultural practices have spoiled the land and water systems upon which food production depended". In spite of this critical world-wide recognition of a deteriorating situation, very little coordinated effort has been put in place to correct the known drawbacks. As a corollary, Water shortage is mostly due to plain mismanagement, at the local and global levels, and those responsible include the governments as well as the private sector and individual consumers. In India, a large number of rural populations live in areas with chronic water shortage and eke out a difficult living. This shortage is more often man made than any authentic lack of the source.

The UN Environmental programme estimates that with increasing population and expanding agriculture, the global demand for water goes up 2 to 3 percent each year. In agrarian based countries like India, similar demand goes up much more steeply than that. However, supply of fresh water remains more or less constant. Naturally, the demand-supply imbalance has an exponential rise with years. This disparity has to be countered by conservation of the available supplies of water, involving prudent water management. The main aspects of water management are protection of water bodies, water-transmitting systems, educating the farmer in scientific uses of water; Government subsidized water sprinkler systems, reducing wastage in the public and personal domain and equalizing distribution of available resources. Major consumer of water is the agricultural sector; most of it in inefficient irrigation. Water used in this sector can be drastically reduced with micro irrigation techniques like drip systems and the more recent subterranean watering of roots that cuts down on the evaporation losses.

In the private sector, and in the personal water use domain, water conservation, despite its critical significance, is one of the most neglected subjects the world over. There are many opportunities to minimize wastage of water for personal use, like closing of taps between uses, promptly fixing leaking pipes, recycling of water for gardening and adopting economical ways of washing vehicles.

Water conservation or saving water is consciously practiced in times of water shortage, but is soon forgotten as soon as sufficient quantities become available. This is particularly discernible in third world countries, where illiteracy plays a large part in depletion of water reserves. In any specific locality, there are large enclaves or communes, where water is available in surplus quantities, where economy in its use is not practiced, while in immediately adjacent areas there exists critical shortage of water even for minimal requirements. To balance and coordinate these two inherently surplus, and innately shortfall areas requires an extraordinary planning by the authorities concerned. A paradigm is a carefully planned distributor network of supply lines, integrated with groundwater from boreholes. Deploying inappropriate pumps in bore and dug wells result in huge wastage of power and water.

Recurring, and justified refrain of the Agricultural Planners, and architects, has been the far-reaching deployment of pumping Units not calibrated to the well capacity, which results in incalculable wastage of power, and water in the energy and irrigation sectors. Apart from such pumps not working to their full efficiency, not fitting power saving devices such as capacitors, appropriate starters, and pump damage preceptors, the lack of system correlation between Head and yield has been causing havoc with the Agricultural and energy investments of the Country. The lack of synchronization between the well, and the pumping system has also been the cause for precipitate failure of an otherwise excellent well. Holding the well culpable for its failure, while actual fault lies with the installation of incorrect pump in the well, leads to erroneous abandonment of the well resulting in wastage of drilling costs and that much groundwater. The answer is to adopt efficient technology and management that can make groundwater exploitation and its use meaningful.

It is very complicated for authorities to exercise any sort of control over exploitation of groundwater. However, it should be feasible to make it mandatory for all drilling agencies, while undertaking drilling, to monitor and provide the following critical data to the User:

1) Volume of water the drill hole yielded during drilling, at various depths, by measuring flow by means of an appropriate devise;
2) The depth to water level in the hole, at every 20m of depth drilled.(by measuring water level by inserting steel tape into the drill rod, while adding rods to the drill string.)

The pump Manufacturers and Dealers should be prohibited from selling a pump to the consumers, unless the 'yield capacity' and 'Head' of the required pump is known and is specified.

Rainwater harvesting:

Groundwater most often is a dependable source of water because of its inherent characteristic of seasonal replenishment and its amenability for long-term storage. Replenishment of groundwater is monsoon dependent; however, deficiency in rainfall for a year or two will not substantially affect the yield from groundwater sources. Problematic reduction in yield may occur, consequent to various extraneous causes, such as deployment of incorrect pumping systems leading to abnormal depletion of water levels and improper protection of the well. Other reasons for reduced yield from wells in urban areas are extensive paving of open areas, which effectively prevents rainwater from percolating into the ground.

Artificial recharge of groundwater, rainwater harvesting, and conservation of groundwater are all synonymous terms. They signify conservation of water, to the extent possible, by not allowing it to flow as drain water, and instead, to improve its infiltration into underground formations, for storage for future abstraction.

Rainwater harvesting aims to induce percolation, and to enhance infiltration of rainwater run-off into the ground by artificial means. It is relatively a new concept envisaging storage and enhancement of groundwater content in certain specific areas involving techniques to accumulate rainwater by various devises in natural underground storage areas and make it available later for use. It is one of the optimum systems available today, and by far the most economical to mitigate the insufficiency of groundwater during summer. If technically designed, structured and implemented appropriately, rainwater harvesting systems play a crucial role in balancing water yield from wells during monsoon and dry seasons. These artificial recharging systems have a fundamental position both in urban and rural water supply schemes.

Rainwater harvesting, in concept and utility, is similar to, and is a miniaturized version of surface reservoirs. Surface reservoirs essentially accumulate and store rainwater draining from respective catchments. Rainwater harvesting structures are built underground, which divert rainwater to enrich groundwater. Being underground, they are not visible, but play an equally important, though a small part, in storing groundwater for later day use.

Systems to draw and store rainwater, for subsequent use are widely prevalent in many of the developed countries in one form or the other. However, in India certain progressive States like Tamil Nadu, Karnataka, Rajasthan, Gujarat and Delhi have adopted rainwater harvesting as a general measure to enhance groundwater infiltration and storage. Provision of simulated recharging structures in all premises is a pre-requisite in many States, but enforcing this stipulation is negligent. This is a consequence of no serious effort having been made to monitor and publicize the extent of advantages accruing from artificially recharging the groundwater reserves. Chennai, a perennially water starved metropolis has been able to prevail over this shortage by strictly enforcing provision of artificial structures in most, if not all buildings. Many States offer 50 to 100% subsidy for measures taken to enhance groundwater recharge. Ministry of water resources, Government of India via their web site 'www.mowr.gov.in' furnish details about procuring or applying for such subsidy. The Center for Science and environment, 41, Tughlakabad Institutional area, New Delhi-110062 (www.cseindia. org) has published a range of books dealing with rainwater harvesting.

Rainwater harvesting pits and trenches

Simplest structures to harvest rainwater essentially consist of 1.5 to 2.5 m wide, and 2.5 to 4 m deep pits with a layer of loose bricks at the bottom. Pits are lined with perforated or loosely cemented bricks and are backfilled with loose boulders, cobbles, gravel, and very coarse sand. It is sealed on top with a concrete mesh or perforated concrete cap. These harvesting pits are excavated in a depressed or level ground to provide easy drainage for the rainwater run-off. The preference for any method is based on the physiographic, geological and soil conditions. In hard rock areas, choice is limited, as in hard rocks groundwater occurs in a dissipated, abstract system restricted to fractures in the formation. Hard massive rocks that have not undergone any weathering, and are devoid of fractures and joints will not hold any water, and will always remain dry.

Around villages, and small settlements, wherever nullas, and minor ephemeral streams flow during monsoons, the monsoon flow is blocked by constructing small earthen bunds with masonry overflow, or a concrete wall with rectangular depressed over flow section. Such structures with average 1 m height check the flow of water. These check 'dams' will let the water spread and stagnate for some time, facilitating infiltration. Ground for such a scheme should be of coarse soil, and physiographically gently sloping from either side towards the nulla with sufficient upstream slope. This artificial spreading of water enhances percolation to enrich the groundwater. Longer and higher the check dam, including the part constructed underground, larger will be the spread. Along gullies and stream courses which experience flooding during the monsoon, wherever the ground profile is favorable i.e., gradually sloping towards the course, number of such check bunds could be built to increase the spreading area. Sloping terrain with 8% surface gradient needs bench terracing for arresting run off, and aiding percolation during monsoons. Such benched terraces could later be used for cultivation.

Wherever the physiography is not suitable for a check dam, trenches are dug across the stream courses and backfilled with puddle clay or any impervious material to arrest underground seepage and create a backflow into the groundwater. Water percolation trenches are typically 30 to 60 cms in width, 2 to 3 m in depth, and extend over lengths of 10 to 20 m, and backfilled with clay also channel water into the ground. In rocky areas, if water courses are not accessible, then the trenches should be excavated across the prevailing fracture system to intercept as many crevices as feasible. Sourcing points are directly located on these trenches.

Percolation tanks:

Excavations made below ground level will let the rain water spread and stagnate for some time, facilitating infiltration. Typical ground for such a scheme should be of coarse soil, and physiographically, gently sloping from either side towards the feeder with sufficient upstream slope. This artificial spreading of water enhances percolation to enrich the groundwater.

Wherever soil, or disintegrated rock (murrum) thickness is of the order of about 20 m, (as can be deduced from dug well excavations, gully sections and similar exposed sections) it is realistic to drill tube wells to feasible depths, near the water spread area, or even along the stream bed to allow rain water to infiltrate underground.

Recharge shafts:

For rural areas, an effective mode of augmenting groundwater resources consists of drilling selected number of wells in dry tank beds, in tanks that fill up only during monsoons, and possess a seasonal transient existence. Number of wells is decided from the dimensions of the tank. Such recharge wells are back filled with cobble, pebble or fresh (unweathered) road metal, which act as filters and as a stabilizer. These wells help infiltration of rain fed waters from the tank into underground formations, and can be drawn from the surrounding wells and hand pump operated wells away from the tank in dry season.

Recharge shafts are similar to small dug wells, of average 1.5 m in diameter are excavated to penetrate the water table. These shafts facilitate rainwater to recharge the groundwater resource during monsoons. Such shafts could be manually excavated if strata are of non-caving material such as murrum. Such shafts or ducts could be drilled by means of calyx drills if the formations are moderately hard, or by reverse rotary drills if the formations are soft. In such a case the shaft sides need to be cased. Unlined shafts need to be back filled with filters consisting of cobbles, pebbles, gravel and coarse sand, beginning with coarse material at the bottom and finer filters near the surface. A short smoothened steining extending above the surface will retard any tendency for the shaft edges to undergo erosion by waters entering the shaft.

Sub-surface dykes:

Sub-surface dykes recharge and store groundwater to their upstream, and constitute good groundwater resources during summer months. These dykes are feasible in physiographically undulating country, in gently sloping valleys of medium width, where bedrock occurs at shallow depths ranging between 5 and 10 m below ground level, overlain by porous valley fills. The chosen site should have a good command area. The dyke is constructed with puddle clay, suitable polythene sheets, mud and bricks, and concrete, depending on local availability. Thickness of the dyke is immaterial, as natural buttresses support it both upstream and downstream. A small diameter jack well upstream of the dyke would form the main intake. The underground dykes stay submerged below valley level and do not impede surface flow during monsoons.

Many rivers that have reached their base level of erosion silt up their own beds, and during summer months tend to flow through such silts, and are visible only in puddles and depressed channels. In such parts of its course, where the river is influent, construction of sub surface dykes are significantly effective in enriching and rising the groundwater levels. Properly constructed sub-surface dykes modify previous effluent streams into influent sources preventing groundwater sources from drying up during dry months.

River beds frequently constitute highly permeable aquifers with good storage potential. Sub-surface structures in the river bed, apart from storing water, back up the ground water to considerable distances both to its upstream and flanks providing water to wells. In areas where water table has declined to low levels, consequent to over exploitation, underground dykes play a significant role in restoring groundwater levels. In physiographically suitable areas, where sub surface dykes are built to extend above the valley fill levels, they contribute to harvesting both surface and sub-surface flows.

Harvesting rainwater from rooftops and terraces:

Recharge pits at suitable intervals, or a single pit is excavated in the available space around the building. Rainwater draining from roof tops and terraces are led through pipes into these pits. Ceramic pipes that lead the water draining from the terraces, to the recharge pits are laid below ground level. Refill filter of a usual recharge pit of 3m depth consists of the following section, but typically, a rainwater harvesting pit should have a depth of 6 meters.

Flush with Ground level	10 to 15 cms thick detachable perforated reinforced concrete slab
0 to 1 m = top layer	45 to 60 cms thick gravel and very coarse sand; not less than 2 mm in size.
1 to 2 m = second layer	45 to 60 cms thick gravel 5 to 10 mm in size
Bottom layer	Boulders of 05 to 15 cms size

Rainwater accumulating in such recharge pits and trenches are conveyed to a dug well or even a tube well or similar groundwater sourcing devices, which in times of inflow will act as infiltration points to recharge the groundwater table.

In Cities and Towns, appropriately designed trenches, excavated adjacent to kerbs, backfilled and sealed on top with perforated concrete will facilitate rainwater percolation, and to a certain extent overcome the dilemma of paved surfaces in Cities preventing rainwater from percolating underground. These kerb side trenches are so placed as to drain the rainwater flooding the roads, and to channel them into the ground.

The preference for any particular method of providing a recharge devise is based on the physiographic, geological and soil conditions. In hard rock areas, choice is limited, as in hard rocks groundwater occurs in a dissipated, abstract system restricted to fractures in the formation. Hard massive rocks that have not undergone any weathering, and are devoid of fractures and joints will not hold any water, and will always remain dry.

Wherever soil, or disintegrated rock (murrum) thickness is of the order of about 20 m, (as can be deduced from dug well excavations, gully sections and similar exposed sections) it is realistic to drill tube wells to feasible depths, near the water spread area, or even along the stream bed to allow rain water to infiltrate underground.

Groundwater levels, rise rapidly and significantly, in regions around reservoirs and canals. This rise increases to a great extent the output capacity from all wells in the area, often calling for replacement of the existing pumps. However, groundwater levels in these areas fluctuate, keeping pace with the rise and fall in the reservoir and canal flow levels.

It is needless to emphasize, that in the interests of both the resident, and the well being of the community at large, a customized artificial rainwater harvesting structure should form an integral part of every building. Additional information, and technical assistance in constructing rainwater harvesting structures could be obtained from the web addresses listed below:

i. www.rainwaterclub.org
ii. www.rainwaterharvesting.org.

These web sites also furnish details of design and construction of rainwater harvesting structures. There are diverse designs appropriate to differing site situations, rainfall conditions, water quantum to be balanced and cost input. Given the design and above factors, Civil Engineers would be able to assemble a custom oriented infiltration structure. The second site also provides addresses of Consultants who provide the required knowledge Bank and skilled workers to execute these structures.

Groundwater resources of India:

Ministry of Water Resources, Government of India, after wide ranging studies has estimated that the annual replenishable groundwater resource of the Country as 433 billion cu meters (1 cu m = 1000 lts = 220.083 imp gallons). This is the assessed quantum of rainwater that infiltrates into a highly averaged ground condition. Water draft, i.e., the groundwater being drawn for use is projected at 231 billion cu m., implying that the quantity that is exploited is about 50% of the available groundwater reserves. Groundwater reserves available for irrigation are approximately 38.284 b cum while the nett draft is 10.65 b cum leaving a balance of 27.635 b cum for further exploitation. It is estimated that there are about twenty million wells in the country, with an irrigation prospect of 46 million hectares, which exceeds the surface water irrigation potential. However, the present status of groundwater development is less than 28% of the available potential. There are many reasons for this under exploitation of a scarce resource. Main reason is the lack of access to command and irrigable areas. The utilization is highly uneven with several regions being constantly over exploited, where groundwater is being overdrawn, in excess of annual replenishment. Greatest challenge today is to balance the groundwater draft between the over exploited and under exploited regions. Repeated overdraft from the reserves has led to lowering of groundwater levels, and in some cases leading to partial drying up of resources. This over exploitation is prevalent particularly in urban sprawls, where, with the rapid expansion of Colonies, habitations have come to depend entirely on groundwater for day-to-day needs. Groundwater is drawn without considering the recharge potential of the area leading to dry wells during the pre-monsoon period. Rainwater harvesting in such areas, wherever adopted, has come to play a vital role in balancing the groundwater reserves.

Future of Indian farming lies, not as much in building dams and reservoirs, as in sustaining well and lift irrigation. This mode of irrigation is easily manageable, and unlike reservoirs does not create problems of land submergence and rehabilitation.

Role of Government Agencies:

The current groundwater development in India is about 58 percent of the utilizable potential. This development is unbalanced, with some areas overdrawing the annual recharge, while some areas have remained dry despite sufficient renewal of the groundwater resources. In India, exploration, development, extraction and use of groundwater is largely a private and individual enterprise. The average farmer who constitutes the majority of this private entrepreneurship, who excavates or drills a well for farming, has no access to the scientific data available or the technology involved in extraction and economic use of groundwater. He depends largely on his own resources, or on some occasions is partly guided by a Consultant. Government agencies dealing with groundwater could play a proactive and positive role in directly facilitating the endeavor of the community in the economic extraction and use of groundwater.

Solar pumps: Pumps operating under solar power are coming into vogue, which proffer the most economic option for rural irrigation needs. Delhi based 'Claro Energy' provides solar power operated irrigation kits with the relevant technical know-how. Initial cost of the Unit may appear high, but is soon balanced by the lesser alternative fuel costs.

BIBLIOGRAPHY AND SELECTED REFERENCES FOR FURTHER READING:

Brinda et al. (Jan 2011). *Fluoride in groundwater, Nalgonda.* Env. assessment vol 172.

Darcy, Henry. (1856). *Les fontaines publiques de la ville de Dijon.* Paris: Delmont.

De Ridder N.A. & Wit K.E. (1965). *Hydraulic cond.of sediments.* Journ.Hydrology vol.3.

Dhruva Narayana et al. (1997). *Watershed management*: ICAR, New Delhi.

Edward E.Johnson Inc.(1981) *Groundwater and wells.* Johnson Divn., UOP Inc., Minnesota.

F.E.Lawrence and P.L.Braunworth, (1906) *Cornell Civil Engineer.*

Felicity Barringer. (Nov 2011). *Ancient water source.* New York Times News Service.

Fenstra L. et al (2007). *Fluoride in groundwater; removal methods*: IGRAC report 1.

Hazen Williams A. *Hydraulics* John Wiley & Sons., New York.

Herman Bouwer. *Groundwater Hydrology.* McGraw Hill Kogakusha Ltd.,

Hubbert M.K. (1940). *The theory of groundwater motion.* Journal of Geology vol 48.

Karanth K.R. *Groundwater assessment, development and management*: McGraw Hill.

Keith E.Anderson. (1981) *Water well Handbook.* Missouri Geological Survey.

Kozisek F. (2004). *Drinking demineralised water.* World Health Organization.
Kruseman, G.P. et al. (1979). *Pumping Test Data.* Inst. for Land Reclamation, Netherlands.
Mahendra A.R. (1963). *Groundwater systems along fault zones*—Guj Jour.of Eng.Res.Ins.
Mahendra A.R. et al (1959). *Groundwater resources of parts of A.P.* Geo. Sur.of Ind
Mahendra A.R. *Shear zones as water conduits.* J, B.Auden commemorative volume ISEG.
Mishra K.K. (2000) *Water the matrix of life.* National Book Trust, India.
Mrunalini V.Khond & Sone A.K. (2012) *Ranney wells* Mining Engineers journal, Vol. 13.
Murthy V.N.S. *Principles of soil mechanics and foundation engineering.* UBS Publishers.
Raghunath H.M. (2002) *Groundwater.* New Age International.
Raju K.C.B. (2012). *Impact of groundwater mining.* Geological Society of India.
Raju K.C.B. et al. (2010). *Coastal salinity prevention.* Third int. conf on Hydrology.
Ralph C Heath and Frank W.Trainer. *Introduction to Groundwater hydrology.* John Wiley.
Ralph C.Heath et al. *Groundwater Hydrology.* John Wiley & Sons.
Ramachandra Rao M.B. *Geophysical prospecting.* University of Mysore.
Sastry G. et al. *Watershed development.* Indian Council of Agricultural research.
Satheesh et al. (July 2004). *Groundwater quality of Chennai City.* Mining Engineers Journal.
Stanley N.Davies and Roger J.M.DeWest. *Hydrogeology.* John Wiley & Sons.
Sudipta Sarkar et al. (May 2005) *Well-head arsenic removal in remote villages of Indian Sub-continent* Water Research. Vol.39. Elsevier. Philidelphia.
Todd David K. (2004). *Groundwater hydrology.* John Wiley & Sons.
Ulli Kulke. (1998). Our *thirsty planet*, New World vol.38.
Volney C.Finch. *Pump Handbook.* Deepak publications, Lucknow.

Wadia D.N. *Geology of India* : Tata McGraw-Hill Publishing Co., Ltd.,
Wilcox L.V. The quality of water for irrigation use. US Dept of Agri. Tech Bulletin.

REPORTS & WEB SITES CITED:

Agricultural pumpsets (1980). Agricultural Refinance & Development Corporation. *www.aqtesolve.com* Aqtesolve software for design and analysis of pumping tests.
Conservation of water (13 Apr 2009). The Times of India.
Evaluation of aquifer parameters (1978). Central Ground water Board, Government of India.
Groundwater over-exploitation committee Report (1979). NABARD. www.geethaborewells.com for large diameter drilling.
Mahendra A.R.et al: *Groundwater exploration of parts of Madras, Punjab, UP.* Tech.Co.Mission; GSI.
Mahendra A.R. et al (1990). *Groundwater investigation for Orient cement plant*: Adilabad.
Manual on *artificial recharge of groundwater.* Central groundwater Board.
*Methods & Techniques of groundwater investigation & developmen*t. UNESCO water resources series No.33.
Minerals in drinking water. Aqua technology.net. (Retrieved).
National seminar on *Geohydrological Scenario.* (2007). Geological Society of India.
Ghosh N.C.& Singh R.D. *ncg@nih.ernet.in.* Nat.Inst.of Hydrology, Roorkee
Open dug wells. Design criteria. Institute of hydraulics & hydrology, Poondi, Tamil Nadu.
Quality control of agricultural pump sets Pilot project studies NABARD.
Report of the *groundwater over-exploitation* committee. NABARD.
Selection of *agricultural pump sets.* Agri. Refinance & Dev. Bank (NABARD).
The development of *groundwater resources.* (2005). UNESCO.
Water distillation—myths and facts. Natural solutions.com. Retrieved
Water resources development—*lift irrigation* : the Institution of Engineers, India.

Water *systems hand book*. Water systems council, Chicago. Water *systems selection guide*. Peabody Barnes, Mansfield, Ohio *Combating waterborne diseases at the household level* (2007). World health Organization

Drinking water standards (2008) World health Organization. Third edition.

World's water resources 30 N0V 2011. "The Hindu"

SUBJECT INDEX

HYDROLOGY

PUMPING SYSTEMS

HORSE POWER

VALVES

INSTALLATION OF PUMPS

PUMP SELECTION

TUBE WELLS

PIPING SYSTEMS

WELLS

MEASURING WATER LEVELS

SAND ANALYSIS

WELL DEVELOPMENT

WELL TEST

MEASURING FLOW

RAINWATER HARVESTING

WATER PURIFICATION

WATER COMPOSITION